Effective Concurrency in Go

Develop, analyze, and troubleshoot high performance
concurrent applications with ease

Burak Serdar

BIRMINGHAM—MUMBAI

Effective Concurrency in Go

Group Product Manager: Gebin George
Publishing Product Manager: Pooja Yadav
Senior Editor: Kinnari Chohan
Technical Editor: Jubit Pincy
Copy Editor: Safis Editing
Project Coordinator: Manisha Singh
Proofreader: Safis Editing
Indexer: Hemangini Bari
Production Designer: Shankar Kalbhor
Developer Relations Marketing Executives: Sonia Chauhan and Rayyan Khan

First published: April 2023

Production reference: 1240323

Published by Packt Publishing Ltd.
Livery Place
35 Livery Street
Birmingham
B3 2PB, UK.

ISBN 978-1-80461-907-0

www.packtpub.com

To Berrin, Selen, and Ersel

– Burak Serdar

Contributors

About the author

Burak Serdar is a software engineer with over 30 years of experience in designing and developing distributed enterprise applications that scale. He's worked for several start-ups and large corporations, including Thomson and Red Hat, as an engineer and technical lead. He's one of the co-founders of Cloud Privacy Labs, where he works on semantic interoperability and privacy technologies for centralized and decentralized systems. Burak holds BSc and MSc degrees in electrical and electronic engineering, and an MSc degree in computer science.

About the reviewer

Tan Quach is an experienced software engineer with a career spanning over 25 years in exotic locations such as London, Canada, Bermuda, and Spain. He has worked with a wide variety of languages and technologies for companies such as Deutsche Bank, Merrill Lynch, and Progress Software and loves diving deep into experimenting with new ones.

Tan's first foray into Go began in 2017 with a proof-of-concept application built over a weekend and productionized and released three weeks later. Since then, Go has been his language of choice when starting any project.

When he can be torn away from the keyboard, Tan enjoys cooking meat over hot coals and open flames and making his own charcuterie boards.

Table of Contents

4

Some Well-Known Concurrency Problems   69

5

Worker Pools and Pipelines   87

6

Error Handling   109

7

Timers and Tickers   121

8

Handling Requests Concurrently   129

9

Atomic Memory Operations   161

10

Troubleshooting Concurrency Issues   171

Index   187

Other Books You May Enjoy   192

Preface

Languages shape the way we think. How we approach problems and formulate solutions for them depends on the concepts we can express using language. This is also true for programming languages. Given a problem, the programs written to solve it may differ from one language to another. This book is about writing programs by expressing concurrent algorithms in the Go language, and about understanding how these programs behave.

Go differs from many popular languages by its emphasis on comprehensibility. This is not the same as readability. Many programs written in easy-to-read languages are not understandable. In the past, I also fell into the trap of writing well-organized programs using frameworks that make programming easy. The problem with that approach is that once writing is over, the program starts a life of its own, and others take over its maintenance. The tribal knowledge that evolved during the development phase is lost, and the team is left with a program that they cannot understand without the help of the last person remaining from the original development team. Developing a program is not that much different from writing a novel. A novel is written so that it can be read by others. So are the programs. If nobody can understand your program, it will not age well.

This book will attempt to explain how to think in the Go language using concurrency constructs so you can understand how the program will behave when you are given a piece of code, and others can understand what you produce. It starts with a high-level overview of concurrency and Go's treatment of it. It will then work on several data processing problems using concurrent algorithms. After all, programs are written to deal with data. I hope that seeing how concurrency patterns develop organically while solving real-life problems can help you acquire the skills to use the language efficiently and effectively. Later chapters will work on more examples involving timing, periodic tasks, server programming, streaming, and practical uses of atomics. The last chapter will talk about troubleshooting, debugging, and additional instrumentation useful for scalability.

It is impossible to cover all topics related to concurrency in a single book. There are many areas left unexplored. However, I am confident that once you work through the examples, you will have more confidence in solving problems using concurrency. Everybody says concurrency is hard. Using the language correctly makes it easier to produce correct programs. The rule of thumb you should always remember is that correctness comes before performance. So, make it work right first, then you can make it work faster.

Who this book is for

If you are a developer who has basic knowledge of the Go language and are looking to gain expertise in highly concurrent backend application development, this is the book for you. This book would also appeal to Go developers of various experience levels in making their backend systems more robust and scalable.

What this book covers

Chapter 1, *Concurrency: A High-Level Overview*, talks about what concurrency is and what it isn't – in particular, how it relates to parallelism. Shared memory and message-passing paradigms, and common concurrency concepts such as race, atomicity, liveness, and deadlock are also introduced in this chapter.

Chapter 2, *Go Concurrency Primitives*, introduces Go language primitives for concurrent programming – namely, goroutines, channels, mutexes, wait groups, and condition variables.

Chapter 3, *The Go Memory Model*, talks about the visibility guarantees of memory operations. It introduces the happened-before relationship that allows you to reason about concurrent behavior, then gives the memory visibility guarantees of concurrency primitives and some of the standard library facilities.

Chapter 4, *Some Well-Known Concurrency Problems*, studies the well-known producer/consumer problem, the dining philosophers problem, and rate-limiting.

Chapter 5, *Worker Pools and Pipelines*, first studies worker pools, which is a common way to process large amounts of data with limited concurrency. Then, it develops several concurrent data pipeline implementations for efficient data processing applications.

Chapter 6, *Error Handling*, explores how to deal with errors and panics in a concurrent program, and how to pass errors around.

Chapter 7, *Timers and Tickers*, shows how to do things periodically and how to do things some time later.

Chapter 8, *Handling Requests Concurrently*, mostly talks about server programming, but many of the concepts discussed in this chapter are broadly about handling requests, so they can be applied in a wide range of scenarios. It describes how to use context effectively, how to distribute work and collect results, how to limit concurrency, and how to stream data.

Chapter 9, *Atomic Memory Operations*, covers atomic memory operations, their memory guarantees, and their practical uses.

Chapter 10, *Troubleshooting Concurrency Issues*, talks about the underrated but essential skill of reading stack traces, and how to detect failures and heal them at runtime.

To get the most out of this book

You need to have a basic understanding of the Go language and a running Go development environment for your operating system. This book does not rely on any other third-party tools or libraries. Use the code editor you are most comfortable with. All examples and code samples can be built and run using the Go build system.

If you are using the digital version of this book, we advise you to type the code yourself or access the code from the book's GitHub repository (a link is available in the next section). Doing so will help you avoid any potential errors related to the copying and pasting of code.

Download the example code files

You can download the example code files for this book from GitHub at `https://github.com/PacktPublishing/Effective-Concurrency-in-Go`. If there's an update to the code, it will be updated in the GitHub repository.

We also have other code bundles from our rich catalog of books and videos available at `https://github.com/PacktPublishing/`. Check them out!

Download the color images

We also provide a PDF file that has color images of the screenshots and diagrams used in this book. You can download it here: `https://packt.link/3rxJ9`.

Conventions used

There are a number of text conventions used throughout this book.

`Code in text`: Indicates code words in text, database table names, folder names, filenames, file extensions, pathnames, dummy URLs, user input, and Twitter handles. Here is an example: "The net/http package implements a Server type that handles each request in a separate goroutine."

A block of code is set as follows:

```
1: chn := make(chan bool) // Create an unbuffered channel
2: go func() {
3:     chn <- true  // Send to channel
4: }()
5: go func() {
6:     var y bool
7:     y <-chn       // Receive from channel
8:     fmt.Println(y)
9: }()
```

Any command-line input or output is written as follows:

```
{"row":65,"height":172.72,"weight":97.61}
{"row":64,"height":195.58,"weight":81.266}
{"row":66,"height":142.24,"weight":101.242}
{"row":68,"height":152.4,"weight":80.358}
{"row":67,"height":162.56,"weight":104.87400000000001}
```

> **Tips or important notes**
> Appear like this.

Get in touch

Feedback from our readers is always welcome.

General feedback: If you have questions about any aspect of this book, email us at customercare@packtpub.com and mention the book title in the subject of your message.

Errata: Although we have taken every care to ensure the accuracy of our content, mistakes do happen. If you have found a mistake in this book, we would be grateful if you would report this to us. Please visit www.packtpub.com/support/errata and fill in the form.

Piracy: If you come across any illegal copies of our works in any form on the internet, we would be grateful if you would provide us with the location address or website name. Please contact us at copyright@packt.com with a link to the material.

If you are interested in becoming an author: If there is a topic that you have expertise in and you are interested in either writing or contributing to a book, please visit authors.packtpub.com.

Share Your Thoughts

Once you've read *Concurrency in Golang*, we'd love to hear your thoughts! Scan the QR code below to go straight to the Amazon review page for this book and share your feedback.

https://packt.link/r/1804619078

Your review is important to us and the tech community and will help us make sure we're delivering excellent quality content.

Download a free PDF copy of this book

Thanks for purchasing this book!

Do you like to read on the go but are unable to carry your print books everywhere? Is your eBook purchase not compatible with the device of your choice?

Don't worry, now with every Packt book you get a DRM-free PDF version of that book at no cost.

Read anywhere, any place, on any device. Search, copy, and paste code from your favorite technical books directly into your application.

The perks don't stop there, you can get exclusive access to discounts, newsletters, and great free content in your inbox daily

Follow these simple steps to get the benefits:

1. Scan the QR code or visit the link below

https://packt.link/free-ebook/9781804619070

2. Submit your proof of purchase
3. That's it! We'll send your free PDF and other benefits to your email directly

1

Concurrency – A High-Level Overview

For many who don't work with concurrent programs (and for some who do), concurrency means the same thing as parallelism. In colloquial speech, people don't usually distinguish between the two. But there are some clear reasons why computer scientists and software engineers make a big deal out of differentiating concurrency and parallelism. This chapter is about what concurrency is (and what it is not) and some of the foundational concepts of concurrency.

Specifically, we'll cover the following main topics in this chapter:

- Concurrency and parallelism
- Shared memory versus message passing
- Atomicity, race, deadlocks, and starvation

By the end of this chapter, you will have a high-level understanding of concurrency and parallelism, basic concurrent programming models, and some of the fundamental concepts of concurrency.

Technical Requirements

This chapter requires some familiarity with the Go language. Some of the examples use goroutines, channels, and mutexes.

Concurrency and parallelism

There was probably a time when concurrency and parallelism meant the same thing in computer science. That time is long gone now. Many people will tell you what concurrency is not: "concurrency is not parallelism," but when it comes to telling what concurrency is, a simple definition is usually elusive. Different definitions of concurrency give different aspects of the concept because concurrency is not how the real world works. The real world works with parallelism. I will try to summarize some of the core ideas behind concurrency, hoping you can understand the abstract nature of it well enough so that you can apply it to solve practical problems.

Many things around us act independently at the same time. There are probably people around you minding their own business, and sometimes, they interact with you and with each other. All these things happen *in parallel*, so parallelism is the natural way of thinking about multiple independent things interacting with each other. If you observe a single person's behavior in a group of people, things are much more *sequential*: that person does things one after the other and may interact with others in the group, all in an orderly sequence. This is quite similar to multiple programs interacting with each other in a distributed system, or multiple threads of a program interacting with each other in a multi-threaded program.

In computer science, it is widely accepted that the study of concurrency started with the work of Edsger Dijkstra – in particular, the one-page paper titled *Solution of a Problem in Concurrent Programming Control* in 1965. This paper deals with a mutual exclusion problem involving N computers sharing memory. This is a very clever description that highlights the difference between concurrency and parallelism: it talks about "concurrent programming" and "parallel execution." **Concurrency relates to how programs are written. Parallelism relates to how programs run.**

Even though this was mostly an academic exercise at the time, the field of concurrency grew over the years and branched into many different but related topics, including hardware design, distributed systems, embedded systems, databases, cloud computing, and more. It is now one of the necessary core skills for every software engineer, thanks to the advances in hardware technology. Nowadays, multi-core processors are the norm, which are essentially multiple processors packed on a single chip, usually sharing memory. These are used in data centers that power cloud-based applications in which someone can provision hundreds of computers connected via a network within minutes, and destroy them after a workload has been completed. The same concurrency principles apply to applications running on multiple machines on a distributed system, to applications running on a multi-core processor in a laptop, and to applications that run on a single-core system. Thus, any serious software developer must be knowledgeable of these principles to develop correct and safe programs that can scale.

Over the years, several mathematical models have been developed to analyze and validate the behavior of concurrent systems. **Communicating Sequential Processes (CSP)** is one such model that influenced the design of Go. In CSP, systems are composed of multiple sequential processes that are running in parallel. These processes can communicate with each other synchronously, which means that a system sending a message to the other can only continue once the other system receives it (this is exactly how unbuffered channels behave in Go.)

The validation aspect of such formal frameworks is most intriguing because they are developed with the promise of proving certain properties of complex systems. These properties can be things such as "can the system deadlock?", which may have life-threatening implications for mission-critical systems. You don't want your auto-pilot software to stop working mid-flight. Most validation activities boil down to proving properties about the states of the program. That's what makes proving properties about concurrent systems so difficult: when multiple systems run together, the possible states of the composite system grow exponentially.

The *state* of a sequential system captures the history of the system at a certain point in time. For a sequential program, the state can be defined as the values in memory together with the current execution location of that program. Given these two, you can determine what the next state will be. As the program executes, it modifies the values of variables and advances the execution location so that the program changes its state. To illustrate this concept, look at the following simple program written in pseudo-code:

```
1: increment x
2: if x<3 goto 1
3: terminate
```

The program starts with `loc=1` and `x=0`. When the statement at location 1 is executed, x becomes 1 and location becomes 2. When the statement at location 2 is executed, x stays the same, but the location goes back to 1. This goes on, incrementing x every time the statement at location 1 runs until x reaches 3. Once x is 3, the program terminates. The sequence in *Figure 1.1* shows the states of the program:

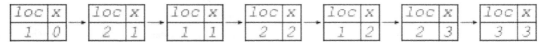

Figure 1.1 – Sequence of the states of the program

When multiple processes are running in parallel, the state of the whole system is the combination of the states of its components. For example, if there are two instances of this program running, then there are two instances of the x variable, which are x_1 and x_2, and two locations, loc_1 and loc_2, pointing to the next line to run. At every state, the possible next states branch based on which copy of the system runs first. *Figure 1.2* illustrates some of the states of this system:

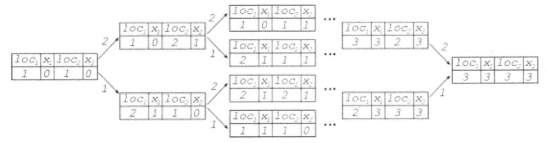

Figure 1.2 – States of the parallel program

In this diagram, the arrows are labeled with the index of the process that runs in that step. A particular run of the composite program is one of the paths in the diagram. Several observations can be made about these diagrams:

- Each sequential process has seven distinct states.

- Each sequential process goes through the same sequence of states at every run, but the states of the two instances of the program interleave in different ways on each path.

- In a particular run, the composite system can go through 14 different states. That is, the length of any path from the start state to the end state in the composite state diagram is 14. (Each process has to go through seven distinct states, making 14 distinct composite states.)

- Every run of the composite system can go through one of the possible paths.

- There are 128 distinct states in the composite system. (For each state of system 1, system 2 can be in 7 distinct states, so $2^7=128$.)

- No matter which path is taken, the end state is the same.

In general, for a system with n states, m copies of that system running in parallel will have n^m distinct states.

That's one of the reasons why it is so hard to analyze concurrent programs: independent components of a concurrent program can run in any order, making it practically impossible to do state analysis.

It is now time to introduce a definition of concurrency:

"Concurrency is the ability of different parts of a program to be executed out-of-order or in partial order without affecting the result."

This is an interesting definition, especially for those who are new to the field of concurrency. For one thing, it does not talk about doing multiple things at the same time, but about executing algorithms "out-of-order." The phrase *doing multiple things at the same time* defines parallelism. Concurrency is about how the program is written, so according to Rob Pike, one of the creators of the Go language, it is about "dealing with multiple things at the same time."

Now, a few words on "ordering" things. There are "total orders," such as the less-than relationship for integers. Given any two integers, you can compare them using the less-than relationship. For sequential programs, we can define a "happened-before relationship," which is a total order: for any two distinct events that happen within a sequential process, one event happens before the other. If two events happen in different processes, how can a happened-before relationship be defined? A globally synchronized clock can be used to order events happening in isolated processes. However, such a clock with sufficient precision does not usually exist in typical distributed systems. Another possibility is causal relationships between processes: if a process sends a message to another when the message is received, anything that happened *before* sending the message *happened before* the second process received it. This is illustrated in *Figure 1.3*:

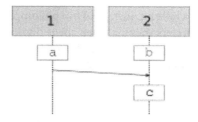

Figure 1.3 – a and b happened before c

Here, event a *happened before* c, and b *happened before* c, but nothing can be said about a and b. They happened "concurrently." In a concurrent program, not every pair of events are comparable, thus the happened-before relationship is a partial order.

Let's revisit the famous "dining philosophers problem" to explore the ideas of concurrency, parallelism, and out-of-order execution. This was first formulated by Dijkstra but later brought to its final form by C.A.R. Hoare. The problem is defined as follows: five philosophers are dining together at the same round table. There are five plates, one in front of each philosopher, and one fork between each plate, five forks total. The dish they are eating from requires them to use both forks, one on their left side, and the other on their right side. Each philosopher thinks for a random interval and then eats for a while. To eat, a philosopher must acquire both forks – one on the left side and one on the right side of the philosopher's plate:

Figure 1.4 – Dining philosophers' problem – some of the possible states

The goal is to devise a concurrent framework that keeps the philosophers well-fed while allowing them to think. We will revisit this problem in detail later as well. For this chapter, we are interested in possible states, some of which are illustrated in *Figure 1.4*. From left to right, the first figure shows all philosophers thinking. The second figure shows two philosophers that picked up the fork on their left-hand side, so one of the philosophers is waiting for the other to finish. The third figure shows the state in which one of the philosophers is eating while the others are thinking. The philosophers next to the one that's eating are waiting for their turn to use the fork. The fourth figure shows the state in which two philosophers are eating at the same time. You can see that this is the maximum number of philosophers that can eat at the same time because there are not enough resources (forks) for one more philosopher to eat. The last figure shows the state where each philosopher has one fork, so they are all waiting to acquire the second fork to eat. This situation will only be resolved if at least one of the philosophers gives up and puts the fork back on to the table so that another one can pick it up.

Now, let's change the problem setup a little bit. Instead of five philosophers sitting at the table, suppose we have a single philosopher who prefers walking when she is thinking. When she gets hungry, she randomly chooses a plate, places the adjacent forks on that plate one by one, and then starts eating. When she is done, she places the forks back on the table one by one and goes back to thinking while walking around the table. She may, however, get distracted during this process and get up at any point, neglecting to put one or both forks back on the table.

When the philosopher chooses a plate, one of the following is possible:

- Both forks are on the table. Then, she picks them up and starts eating.

- One of the forks is on the table, and the other one is on the next plate. Realizing that she cannot eat with a single fork, she gets up and chooses another plate. She may or may not put the fork back on the table.

- One of the forks is on the table, and the other one is on the selected plate. She picks up the second fork and starts eating.

- None of the forks are on the table, because they are both on adjacent plates. Realizing that she cannot eat without a fork, she gets up and chooses another plate.

- Both forks are on the selected plate. She starts eating.

Even though the modified problem has only one philosopher, the possible states of the modified problem are identical to those of the original. The five states depicted in the preceding figure are still some of the possible states of the modified problem. The original problem, where there are five processors (philosophers) performing a computation (eating and thinking) using shared resources (forks) illustrates the parallel execution of a concurrent program. In the modified program, there is only one processor (philosopher) performing the same computation using shared resources by dividing her time (time sharing) to fulfill the roles of the missing philosophers. The underlying algorithms (behavior of the philosopher(s)) are the same. So, **concurrent programming is about organizing a problem into computational units that can run using time sharing or that can run in parallel**. In that sense, concurrency is a programming model like object-oriented programming or functional programming. Object-oriented programming divides a problem into logically related structural components that interact with each other. Functional programming divides a problem into functional components that call each other. Concurrent programming divides a problem into temporal components that send messages to each other, and that can be interleaved or run in parallel.

Time-sharing means sharing a computing resource with multiple users or processes. In concurrent programming, the shared resource is the processor itself. When multiple threads of executions are created by a program, the processor runs one thread for some time, and then switches to another thread, and so on. This is called context-switching. The context of an execution thread contains its stack and the states of the processor registers when that thread stopped. This way, the processor can quickly switch from stack to stack, saving and restoring the processor's state at each switch. The exact location in the code where the processor does that switch depends on the underlying implementation. In **preemptive threading**, a running thread can be stopped at any time during that thread's execution.

In **non-preemptive threading** (or **cooperative threading**), a running thread voluntarily gives up execution by performing a blocking operation, a system call, or something else.

For a long time (until Go version 1.14), the Go runtime used a cooperative scheduler. That meant that in the following program, once the first goroutine started running, there was no way to stop it. If you build this program with a Go version < 1.14 and run it with a single OS thread multiple times, some runs will print Hello, while others will not. This is because if the first goroutine starts working before the second one, it will never let the second goroutine run:

```go
func main() {
    ch:=make(chan bool)
    go func() {
        for {}
    }()
    go func() {
        fmt.Println("hello")
    }()
    <-ch
}
```

This is no longer the case for more recent Go versions. Now, the Go runtime uses a preemptive scheduler that can run other goroutines, even if one of them is trying to consume all processor cycles.

As a developer of concurrent systems, you have to be aware of how threads/goroutines are scheduled. This understanding is the key to identifying the possible ways in which a concurrent system can behave. At a high level, the states a thread/goroutine can be in are shown using the state in *Figure 1.5*:

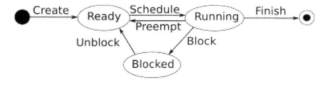

Figure 1.5 – Thread state diagram

When a thread is created, it is in the *Ready* state. When the scheduler assigns it to a processor it moves to the *Running* state and starts running. A running thread can be preempted and moved back into the *Ready* state. When the thread performs an I/O operation or blocks waiting for a lock or channel operation, it moves to the *Blocked* state. When the I/O operation completes, the lock is unlocked, or the channel operation is completed, the thread moves back to the *Ready* state, waiting to be scheduled.

The first thing you should notice here is that a thread waiting for something to happen in a blocked state may not immediately start running when it is unblocked. This fact is usually overlooked when designing and analyzing concurrent programs. What is the meaning of this for your Go programs? It means that unlocking a mutex doesn't mean one of the goroutines waiting for that mutex will start running immediately. Similarly, writing to a channel does not mean the receiving goroutine will immediately start running. They will be ready to run, but they may not be scheduled immediately.

You will see different variations of this thread state diagram. Every operating system and every language runtime has different ways of scheduling their execution threads. For example, a threading system may differentiate between being blocked by an I/O operation and being blocked by a mutex. This is only a high-level depiction almost all thread implementations share.

Shared memory versus message passing

If you have been developing with Go for some time, you have probably heard the phrase "Do not communicate by sharing memory. Instead, share memory by communicating." Sharing memory among the concurrent blocks of a program creates vast opportunities for subtle bugs that are hard to diagnose. These problems manifest themselves randomly, usually under load that cannot be simulated in a controlled test environment, and they are hard or impossible to reproduce. What cannot be reproduced cannot be tested, so finding such problems is usually a matter of luck. Once found, they are usually easy to fix with very minor changes. That adds insult to injury. Go supports both shared memory and message-passing models, so we will spend some time looking at what the shared memory and message-passing paradigms are.

In a shared memory system, there can be multiple processors or cores with multiple execution threads that use the same memory. In a **Uniform Memory Access (UMA)** system, all processors have equal access to memory. This can reduce throughput because the same memory access bus is shared by multiple processors. In a **Non-Uniform Memory Access (NUMA)** system, processors may have dedicated access to certain blocks of memory. In such a system, the operating system can allocate the memory of the processes that run in a processor to the memory region dedicated to that processor, increasing the memory throughput. Almost all systems also use faster cache memory to improve memory access time, and they employ cache coherence protocols to make sure that processes do not read stale values (values that are updated in the cache but not written to the main memory) from the main memory.

Within a single program, shared memory simply means that multiple execution threads access the same part of memory. Writing programs using shared memory comes naturally in most situations. In most modern languages such as Go, execution threads have unrestricted access to the memory space of the whole process. So, any variable accessible to a thread can be read and modified by that thread. If there were no compiler or hardware optimizations, this would have been just fine. When executing programs, modern processors do not usually wait for their memory-read or memory-write operations to complete. They execute instructions in a pipeline. So, while an instruction is waiting for the completion of a read/write operation, the processor can start running the next instruction of the program. If the second instruction completes before the first one, then the results of that instruction may be written to memory "out-of-order."

Let's look at this very simple program as an example:

```
var locked,y int
func f() {
    locked=1
    y=2
    ...
}
```

In this program, suppose the f() function starts with locked and when y is initialized to 0. Then, locked is set to 1, presumably to implement some sort of a locking scheme, and then y is set to 2. If we take a snapshot of the memory at that point, we may see one of the following:

- locked=0, y=0: The processor ran the assignment operations, but the updates are not written to memory yet

- locked=0, y=2: The y variable has been updated in memory, but locked is not yet updated in memory

- locked=1, y=0: The locked variable is updated and written to memory, but the y variable may or may not be updated yet

- locked=1, y=2: Both variables are updated, and the updates are written to memory

Reordering instructions is not limited to processors. Compilers can also move statements around without this affecting the result of computation within a sequential program. This is usually done during the optimization phase of the compilation. In the preceding program, nothing is preventing the compiler from reversing the order of the two assignments. Based on the rest of the code, the compiler may decide that executing locked=1 after y=2 is better.

In short, there is no guarantee for one thread to see the correct values of variables modified by other threads without additional mechanisms.

To make shared memory applications work correctly, we need a way to tell the compiler and the processor to commit all changes to memory. The low-level facility that is used for that purpose is a **memory barrier**. A memory barrier is an instruction that forces a certain ordering constraint on the processor and the compiler. After a memory barrier, all operations issued before the barrier are guaranteed to be completed before those that come after the barrier. In the preceding program, a memory barrier after the assignment to y ensures that the snapshot will have locked=1 and y=2. So, a memory barrier is a crucial low-level feature we need to make shared memory applications run correctly.

You may be wondering, how is this useful for me? When you are dealing with concurrent programs that use shared memory, the effects of the operations performed in one block will only be guaranteed to be visible to the other concurrent blocks after a memory barrier. These are certain operations in Go: atomic read/writes (functions in sync/atomic package) and channel read/writes. Other synchronization operations (mutexes, wait groups, and condition variables) use the atomic read/writes and channel operations, and thus also include memory barriers.

> **Note**
>
> A quick note about debugging concurrent programs is necessary here. A common practice for debugging concurrent programs is to print the values of some variables at a critical point in code to capture information about the state of the program when some unexpected behavior occurs. This practice usually prevents that unexpected behavior, because printing something to the console or writing a log message to a file usually involves mutexes and atomic operations. So, it is more common to have bugs in production when logging is turned off than to have them during development where there is much logging.

Now, a few words about the message-passing model. For many languages, concurrency is an add-on feature with libraries defining functions to create and manage concurrent blocks (threads) of execution. Go takes a different approach by making concurrency a core feature of the language. The concurrency features of Go are inspired by CSP, where multiple isolated processes communicate by sending/receiving messages. This has also been the basic process model for Unix/Linux systems for a long time. The Unix motto has been "make each program do one thing well," and then compose these programs so that they can do more complicated tasks. Most traditional Unix/Linux programs are written to read/ write from/to the terminal so that they can be connected one after the other in the form of a pipeline, where each program uses the output of the previous one as input using interprocess communication mechanisms. This system also resembles a distributed system with multiple interconnected computers. Each computer has dedicated memory where it carries out computations, sends its results to other computers on the network, and receives results from other computers.

Message passing is one of the core ideas in establishing happened-before relationships in distributed systems. When a system receives a message from another system, you can be sure that any event that happened before the sender system sent that message *happened before* the receiving system received it. Without such causal references, it is usually not possible to identify which event happened when in a distributed system. The same idea applies to concurrent systems. In a message-passing system, happened-before relationships are established using the messages. In shared memory systems, such happened-before relationships are established using locks.

There are some advantages to having both models available – for instance, many algorithms are much easier to implement on a shared memory system, whereas a message-passing system is free of certain types of data races. There are some disadvantages as well. For instance, in a hybrid model, it is easy to share memory unintentionally, creating data races. A common theme to prevent such unintentional sharing is to be mindful of data ownership: when you send a data object to another goroutine via a channel, the ownership of that data object is transferred to the receiving goroutine, and the sending goroutine must not access that object again without ensuring mutual exclusion. Sometimes, this is hard to achieve.

For example, even though the following snippet uses a channel to communicate between threads, the object sent via the channel is a map, which is a pointer to a complicated data structure. Continuing to use the same map after the channel send operation will probably result in data corruption and panic:

```
// Compute some result
data:=computeData()
m:=make(map[string]Data)
// Put the result in a map
m["result"]=data
// Send the map to another goroutine
c<-m
// Here, m is shared between the other goroutine and this one
```

Atomicity, race, deadlocks, and starvation

To write and analyze concurrent programs successfully, you have to be aware of some key concepts: atomicity, race, deadlocks, and starvation. Atomicity is a property you have to carefully exploit for safe and correct operation. Race is a natural condition related to the timing of events in a concurrent system, and can create irreproducible subtle bugs. You have to avoid deadlocks at all costs. Starvation is usually related to scheduling algorithms, but can also be caused by bugs in the program.

A **race condition** is a condition in which the outcome of a program depends on the sequence or timing of concurrent executions. A race condition is a bug when at least one of the outcomes is undesirable. Consider the following data type representing a bank account:

```
type Account struct {
     Balance int
}

func (acct *Account) Withdraw(amt int) error {
     if acct.Balance < amt {
          return ErrInsufficient
     }
     acct.Balance -= amt
     return nil
}
```

The `Account` type has a `Withdraw` method that checks the balance to see if there are sufficient funds for withdrawal, then will either fail with an error or reduce the balance. Now, let's call this method from two goroutines:

```
acct:=Account{Balance:10}
go func() {acct.Withdraw(6)}() // Goroutine 1
go func() {acct.Withdraw(5)}() // Goroutine 2
```

The logic should not allow the account balance to go below zero. One of the goroutines should run successfully, while the other one should fail because of insufficient funds. Depending on which goroutine runs first, the account should have either 5 or 4 for the final balance. However, these goroutines may interleave, resulting in many possible runs, some of which are shown here:

Possible run 1		
acct.Balance=10		
Goroutine 1 (amt=6)	Goroutine 2 (amt=5)	acct.Balance
if acct.Balance < amt		10
	if acct.Balance < amt	10
	acct.Balance -= amt	5
acct.Balance -= amt		-1
Possible run 2		
acct.Balance=10		
Goroutine 1 (amt=6)	Goroutine 2 (amt=5)	acct.Balance
	if acct.Balance < amt	10
if acct.Balance < amt		10
acct.Balance -= amt		4
	acct.Balance -= amt	-1
Possible run 3		
acct.Balance=10		
Goroutine 1 (amt=6)	Goroutine 2 (amt=5)	acct.Balance
	if acct.Balance < amt	10
	acct.Balance -= amt	5
if acct.Balance < amt		5
return ErrInsufficient		5

Possible run 4		
acct.Balance=10		
Goroutine 1 (amt=6)	Goroutine 2 (amt=5)	acct.Balance
`if acct.Balance < amt`		10
`acct.Balance -= amt`		4
	`if acct.Balance < amt`	4
	`return ErrInsufficient`	4

Possible runs 1 and 2 leave the account with a negative balance, even though the logic prevents this case for a single-threaded program. Possible runs 3 and 4 are also race conditions but they leave the system in a consistent state. Two goroutines compete for the funds; one of them is successful while the other fails. This makes race conditions very hard to diagnose: they only happen rarely, and they are not reproducible, even if the conditions under which they are detected are replicated.

The race condition shown here is also a **data race**. A data race is a special type of race condition where the following occurs:

- Two or more threads are accessing the same memory location
- At least one of the threads writes to that location
- There is no synchronization or locks to coordinate the ordering of operations between the two threads

Note that in runs 3 and 4, first, the function ran in its entirety in one goroutine, then it ran in its entirety in the other goroutine. Thus, Withdraw ran *atomically*. An **atomic operation** contains a sequence of sub-operations where either none of them happen, or all of them happen.

Now, let's consider a more realistic scenario: the effects of the sequential operations of one goroutine may look out-of-order to another goroutine. The following is a possible run:

Possible run 5			
acct.Balance=10			
Goroutine 1 (amt=6)	Balance	Goroutine 2 (amt=5)	Balance
	10	`if acct.Balance < amt`	10
	10	`acct.Balance -= n`	5
`if acct.Balance< n`	10		5
`acct.Balance-= n`	4		4

In run 5, the compiler and the processor provide a consistent sequential view of the memory for each goroutine, which may not be the same for all goroutines. In the preceding example, the memory-write operation (acct.Balance -= n) by goroutine 2 is delayed, resulting in goroutine 1 deciding there are enough funds. Eventually, goroutine 2 observes that the update made by goroutine 1 was written to memory, and the update made by goroutine 1 is lost.

Such data races are possible in any shared memory program where multiple execution threads can access the same object concurrently. If the operations changing the contents of the shared object are performed atomically, then they appear to be instantaneous. One way to implement atomicity is by using **critical sections**. A critical section is a protected section of a program in which only one process or thread can make modifications to a shared resource. When only one thread is allowed to enter the critical section, the **mutual exclusion** property is satisfied, and the operation that's performed in the critical section is atomic. A subtle point we need to emphasize here is that a critical section is defined *for a specific shared resource*. Multiple processes can be in their critical sections and satisfy mutual exclusion property if they do not share a resource.

A sync.Mutex can be used for mutual exclusion:

```
type Account struct {
    sync.Mutex
    ID string
    Balance int
}


func (acct *Account) Withdraw(amt int) error {
    acct.Lock()
    defer acct.Unlock()
    if acct.Balance < amt {
        return ErrInsufficientFunds
    }
    acct.Balance -= amt
    return nil
}
```

Let's analyze how this program works: one of the goroutines arrives at acct.Lock(), attempts to lock the mutex, and succeeds. If any other goroutine arrives at the same point, their attempts to lock the mutex will be blocked until the first goroutine unlocks it. The first goroutine, now that it has entered the critical section successfully, completes the function and unlocks the mutex. The unlocking operation enables all the other goroutines that are waiting for it to become unlocked. One of the waiting goroutines is randomly selected, so it locks the mutex and takes its turn running the critical section. The mutex implementation contains a memory barrier, so all the modifications that are performed by the first goroutine are visible to the second goroutine. An important point to note here is the use of the **shared** `acct.Mutex`. Atomicity is only guaranteed if all access to `acct.Balance` is done when the shared mutex is locked. The second thread sees the effect of the first thread's call to `Withdraw()` as an instantaneous execution because the second thread is blocked while waiting for the first thread.

You may have already noticed that all concurrent operations using a read-decide-update scheme are prone to race conditions unless the read-decide-update block is in a critical section. For example, if you are working with a database, reading an object and then updating it with another database invocation has a race condition. This is because, after reading the object, another process may change it before we can write it back. If you are dealing with a file system, creating a file by checking if it exists first is a race condition because another process may create the file right after you check its existence but before you create it, causing you to overwrite an existing file. If you are working with a variable in memory, another thread may modify the variable after one thread reads it, causing it to overwrite potential changes. Most systems come with a way to define critical sections. Databases have transaction management systems to ensure atomicity and mutual exclusion; shared memory systems have mutexes.

The use of mutexes can get complicated when multiple shared objects need to interact. Let's say we have two accounts, and we want to write a function to transfer money from one to the other. The operation has to be atomic to ensure the total amount of money in the system is consistent. The following is our first attempt:

```
func Transfer(from, to *Account, amt int) error {
    from.Lock()
    defer from.Unlock()
    to.Lock()
    defer to.Unlock()
    if from.Balance < amt {
        return ErrInsufficient
    }
    from.Balance -= amt
    to.Balance += amt
}
```

As we did previously, we call this function from two goroutines:

```
acct1 := Account{Balance: 10}
acct2 := Account{Balance: 15}
go func() { Transfer(acct1, acct2, 5) }()
go func() { Transfer(acct2, acct1, 10) }()
```

The following is a possible execution of this function:

Possible run 1	
Goroutine 1 (from acct1 to acct2)	Goroutine 2 (from acct2 to acct1)
`acct1.Lock()`	
	`acct2.Lock()`
`acct2.Lock()// blocked`	
	`acct1.Lock() // blocked`

After locking `acct1`, goroutine 1 attempts to lock `acct2`, which was locked by goroutine 2 before. Goroutine 1 blocks this. However, goroutine 2 attempts to lock `acct1`, and blocks because `acct1` is locked by goroutine 1. This is a **deadlock**. A deadlock is a situation in which every member of a group of objects is waiting on objects in the same group to release a lock. There are four necessary and sufficient conditions for a deadlock to happen (these are called Coffman conditions):

1. At least one member must hold a resource exclusively.

2. At least one member must be waiting for a resource being held by another member while holding another one exclusively.

3. Only the member holding a resource can release that resource.

4. There is a set of members, P_1, P_2, ..., P_n, such that P_1 is waiting for a resource held by P_2, P_2 is waiting for a resource held by P_3, ..., and P_n is waiting for a resource held by P_1.

It is usually not possible to change the resource requirements of the algorithm: such a transfer function will need to lock both accounts. So, one way to fix this deadlock problem is to remove the fourth condition: the cyclic wait. Here, goroutine 1 is waiting for a lock held by goroutine 2, while goroutine 2 is waiting for a lock held by goroutine 1. A simple way to resolve this is to impose a consistent ordering of locks whenever multiple objects must be locked. The `Account` object has a unique identifier, so we can order the locking of accounts by their identifiers:

```
func Transfer(from, to *Account, amt int) error {
    if from.ID < to.ID {
        from.Lock()
        defer from.Lock()
        to.Lock()
        defer to.Lock()
    } else {
        to.Lock()
        defer to.Lock()
        from.Lock()
```

```
                defer from.Lock()
    }
...
}
```

This is not always easy, especially if the algorithm involves conditional locking of objects. Enforcing a consistent order of locking for multiple resources usually prevents many deadlocks, but not all.

As a side note, the Go runtime uses the term "deadlock" more liberally. If the runtime detects that all goroutines are blocked, it calls it a deadlock. This is quite easy to detect by the scheduler: it knows all goroutines and which ones are active, so it can schedule them. If there is none, it panics. For example, none of the four conditions of deadlock apply to the following program, but the Go runtime calls it a deadlock:

```
func main() {
    c := make(chan bool)
    <-c
}
```

The Go runtime does not detect deadlocks that occur in a proper subset of the goroutines. That's the case if some goroutines are deadlocked in a cyclic wait while other goroutines are still running. The deadlocked goroutines will continue waiting until the program ends. This can be a critical problem for long-running programs such as servers.

Starvation is a situation in which a process is denied access to a resource for a long time or indefinitely. It is usually a result of simplistic or incorrect scheduling algorithms and bugs in algorithms. For example, let's consider two goroutines where the first one locks a mutex for a long time, and the second one locks the same mutex for a short time:

```
var mutex sync.Mutex
    go func() {
        for {
            mutex.Lock()
            // Work for a long time
            mutex.Unlock()
            // Do something quick
        }
    }()
    go func() {
        for {
            mutex.Lock()
```

```
            // Do something quick
            mutex.Unlock()
            // Work for a long time
        }
    }()
```

The first goroutine keeps the mutex locked for a long time while the second goroutine is waiting for it. When it releases the mutex, both goroutines will be competing for the mutex again. A schedule that gives the goroutines equal chance will select the first goroutine half of the time. Also, whenever the second goroutine releases the lock, it is almost certain that the first goroutine will be waiting to acquire it because the second goroutine has a long task to complete. So, a simple "fair" scheduler would unfairly "starve" the second goroutine. Recent versions of the Go runtime deal with this problem by favoring goroutines waiting in the queue, thus giving both goroutines a fair chance.

Starvation is also the key idea for denial of service attacks. The idea of a denial of service attack is to starve a computer with useless computations so much so that it cannot perform any useful computations. For example, a web service may accept an XML document containing a recursive entity definition and parse it, resulting in an XML document with billions of entries (this is called a Billion Laughs Attack, which generates a billion "lol"s when a small XML document is processed). As a rule of thumb for services, never trust the data you receive from a client. Always impose size limits and do not blindly expand the data you receive.

The Go runtime helps detect deadlocks where all goroutines are asleep, but it fails to detect situations where goroutines are starved. Such situations can be difficult to detect. Take a look at the following example:

```
func Producer() <-chan string {
    c := make(chan string)
    go func() {
        defer func() {
            if r := recover(); r != nil {
                log.Println("Recovered", r)
            }
        }()
        for i := 0; i < 10; i++ {
            c <- produceValue()
        }
        close(c)
    }()
    return c
}
```

```
func httpHandler(w http.ResponseWriter, req *http.Request) {
    c := Producer()
    for _, x := range c {
        w.Write([]byte(x))
    }
}
```

This is a web service handler that responds to a request by returning the strings produced by a producer function. When the request comes, the handler creates a `Producer` and expects to receive some values from it through a channel. When the producer is done generating values, it closes the channel so that the `for` loop can terminate. Now, suppose the `produceValue` function panics. That will terminate the goroutine without closing the channel. Panics are propagated through the stack of the goroutine it happened in. The HTTP handler, running in a separate goroutine, never receives that panic and is now indefinitely blocked. The client who issued the request may eventually timeout, but the goroutine created for the HTTP handler will never terminate (the goroutine will **leak**). This will not be detected by the runtime, though it may be observed in the logs of this server. If this panic happens sufficiently often, the server will soon run out of resources to continue. (This example has an easy fix: instead of closing the channel at the end of the goroutine, close it in a defer statement at the beginning. That will close the channel, even in case of panic.)

A **livelock** is a situation that looks like a deadlock, but none of the processes are blocked waiting on another. It is a kind of starvation in the sense that even though processes work, they don't do anything useful. A typical example of a livelock is two people walking in opposite directions in a narrow hallway. When they meet, both being equally polite, each wants to yield to the other, but they both make the same decisions at the same time, moving left and right in synchrony while failing to clear the way for the other. The runtime cannot offer much help for a livelock because as far as the runtime is concerned, the processes are running. In a livelock situation, processes keep attempting to acquire a resource and fail because the other process is also doing the same. It is easy to fall into a livelock using a try-lock:

```
go func() {
    for {
        if mutex1.TryLock() {
            if mutex2.TryLock() {
                // ...
                mutex2.Unlock()
            }
            mutex1.Unlock()
        }
    }
}
```

```
    }()
go func() {
    for {
        if mutex2.TryLock() {
            if mutex1.TryLock() {
                // ...
                mutex1.Unlock()
            }
            mutex2.Unlock()
        }
    }
}()
```

A livelock is very difficult to diagnose because the situation may resolve itself after several iterations. There are several ways you can prevent a livelock: if you fail to acquire a resource and intend to retry it (such as by using `TryLock`), retry for a finite number of times, and possibly with random delays between retries. This can be generalized to work queues and pipelines: if you acquire a piece of work and fail to complete it, record the number of times that piece of work has been retried so that you don't keep rescheduling the same work indefinitely.

These concepts can be used to define a framework for specifying certain properties of concurrent programs. There are usually two kinds of properties of interest:

- Safety properties specify that something bad never happens. The absence of a deadlock is an important safety property that states no execution of a program can deadlock. Mutual exclusion guarantees that no two processes sharing a resource are in their critical sections at the same time. Correctness properties guarantee that if a program starts at a valid state, it should end at a valid state.

- Liveness properties specify that something good eventually happens. One of the most studied liveness properties is that the program should eventually terminate. This does not usually hold for long-running programs such as servers. For such programs, important liveness properties include that every request must eventually be answered, a message sent will eventually be received, and a process will eventually enter its critical section.

There have been many studies on providing safety and liveness properties. For this book, our focus will be on practical applications of such properties. For example, the liveness of applications can usually be tested by heartbeat services. This can detect if the application became completely unresponsive, but cannot detect if part of the application is busy spinning in a livelock. For some mission-critical software systems, formal proof of safety and liveness properties is necessary. For many others, runtime evaluation and detection of these properties is sufficient. With careful observation of the internal workings of a concurrent program, it is usually possible to test certain safety and liveness properties.

A safety or liveness problem can sometimes be remedied by terminating that component and creating another instance of it. This approach, in a way, gives up on the idea that all such problems must be identified before deploying the system and accepts that the programs will fail. What is important is that when programs fail, they should be recognized quickly, and fixed by automatically replacing the component. This is one of the core tenets of cloud computing and the microservice architecture: individual components will fail but the overall system must self-heal by detecting and replacing failing components automatically.

Summary

The main theme in this chapter was that concurrency is not parallelism. Parallelism is an intuitive concept people are used to because the real world works in parallel. Concurrency is a mode of computation where blocks of code may or may not run in parallel. The key here is to make sure we get the correct result no matter how the program is run.

We also talked about the two main concurrency programming paradigms: message passing and shared memory. Go permits both, which makes it easy to program, but equally easy to make mistakes. The last part of this chapter was about fundamental concepts of concurrent programming – that is, race conditions, atomicity, deadlocks, and livelock concepts. The important point to note here is that these are not theoretical concepts – these are real situations that affect how programs run and how they fail.

We tried to avoid Go specifics in this chapter as much as possible. The next chapter will cover Go concurrency primitives.

Question

We looked at the dining philosopher's problem with a single philosopher that walks when they are thinking. What problems can you foresee if there are two philosophers?

Further reading

The literature on concurrency is very rich. These are only some of the seminal works in the field of concurrency and distributed computing that are related to the topics we discussed in this chapter. Every serious software practitioner should at least have a basic understanding of these.

The following paper is easy to read and short. It defines mutual exclusion and critical sections: *E. W. Dijkstra. 1965. Solution of a problem in concurrent programming control. Commun. ACM 8, 9 (Sept. 1965), 569.* https://doi.org/10.1145/365559.365617.

This is the CSP book. It defines the CSP as a formal language: *Hoare, C. A. R. (2004) [originally published in 1985 by Prentice Hall International]. "Communicating Sequential Processes" (PDF). Usingcsp.com.*

The following paper talks about the ordering of events in a distributed system: *Time, Clocks and the Ordering of Events in a Distributed System, Leslie Lamport, Communications of the ACM 21, 7 (July 1978), 558-565. Reprinted in several collections, including Distributed Computing: Concepts and Implementations, McEntire et al., ed. IEEE Press, 1984. | July 1978, pp. 558-565. 2000 PODC Influential Paper Award (later renamed the Edsger W. Dijkstra Prize in Distributed Computing). Also awarded an ACM SIGOPS Hall of Fame Award in 2007.*

2

Go Concurrency Primitives

This chapter is about the fundamental concurrency facilities of the Go language. We will first talk about goroutines and channels, the two concurrency building blocks that are defined by the language. Then, we also look at some of the concurrency utilities that are included in the standard library. We will cover the following topics:

- Goroutines
- Channels and the `select` statement
- Mutexes
- Wait groups
- Condition variables

By the end of this chapter, you will have enough under your belt to tackle basic concurrency problems using language features and standard library objects.

Technical Requirements

The source code for this particular chapter is available on GitHub at `https://github.com/PacktPublishing/Effective-Concurrency-in-Go/tree/main/chapter2`.

Goroutines

First, some basics.

A **process** is an instance of a program with certain dedicated resources, such as memory space, processor time, file handles (for example, most processes in Linux have `stdin`, `stdout`, and `stderr`), and at least one thread. We call it an instance because the same program can be used to create many processes. In most general-purpose operating systems, every process is isolated from the others, so any two processes that wish to communicate have to do it through well-defined inter-process communication utilities. When a process terminates, all the memory allocated for the process is freed, all open files are closed, and all threads are terminated.

A **thread** is an execution context that contains all the resources required to run a sequence of instructions. Usually, this contains a stack and the values of processor registers. The stack is necessary to keep the sequence of nested function calls within that thread, as well as to store values declared in the functions executing in that thread. A given function may execute in many different threads, so the local variables used when that function runs in a thread are stored in the stack of that thread. A **scheduler** allocates processor time to threads. Some schedulers are preemptive and can stop a thread at any time to switch to another thread. Some schedulers are collaborative and have to wait for the thread to yield to switch to another one. A thread is usually managed by the operating system.

A **goroutine** is an execution context that is managed by the Go runtime (as opposed to a thread that is managed by the operating system). A goroutine usually has a much smaller startup overhead than an operating system thread. A goroutine starts with a small stack that grows as needed. Creating new goroutines is faster and cheaper than creating operation system threads. The Go scheduler assigns operating system threads to run goroutines.

In a Go program, goroutines are created using the go keyword followed by a function call:

```
go f()
go g(i,j)
go func() {
...
}()
go func(i,j int) {
...
}(1,2)
```

The go keyword starts the given function in a new goroutine. The existing goroutine continues running concurrently with the newly created goroutine. The function running as a goroutine can take parameters, but it cannot return a value. The parameters of the goroutine function are evaluated before the goroutine starts and passed to the function once the goroutine starts running.

You may ask why there was a need to develop a completely new threading system. Just to get lightweight threads? Goroutines are more than just lightweight threads. They are the key to increasing throughput by efficiently sharing processing power among goroutines that are ready to run. Here's the gist of the idea.

The number of operating system threads used by the Go runtime is equal to the number of processors/ cores on the platform (unless you change this by setting the GOMAXPROCS environment variable or by calling the runtime.GOMAXPROCS function). This is the number of things the platform can do *in parallel*. Anything more than that and the operating system will have to resort to time sharing. With GOMAXPROCS threads running in parallel, there is no context-switching overhead at the operating system level. The Go scheduler assigns goroutines to operating system threads to get more work on each thread as opposed to doing less work on many threads. The smaller context switching is not the only reason why the Go scheduler performs better than the operating system scheduler. The Go scheduler performs better because it knows which goroutines to wake up to get more out of them. The operating system does not know about channel operations or mutexes, which are all managed in the user space by the Go runtime.

There are some more subtle differences between threads and goroutines. Threads usually have priorities. When a low-priority thread competes with a high-priority thread for a shared resource, the high-priority thread has a better chance of getting it. Goroutines do not have pre-assigned priorities. That said, the language specification allows for a scheduler that favors certain goroutines. For example, later versions of the Go runtime include scheduling algorithms that will select starving goroutines. In general, though, a correct concurrent Go program should not rely on scheduling behavior. Many languages have facilities such as thread pools with configurable scheduling algorithms. These facilities are developed based on the assumption that thread creation is an expensive operation, which is not the case for Go. Another difference is how goroutines stacks are managed. A goroutine starts with a small stack (Go runtimes after 1.19 use a historical average, earlier versions use 2K), and every function call checks whether the remaining stack space is sufficient. If not, the stack is resized. An operation system thread usually starts with a much larger stack (in the order of megabytes) that usually does not grow.

The Go runtime starts several goroutines when a program starts. Exactly how many depends on the implementation and may change between versions. However, there is at least one for the garbage collector and another for the main goroutine. The main goroutine simply calls the `main` function and terminates the program when it returns. When `main` returns and the program exits, all running goroutines terminate abruptly, mid-function, without a chance to perform any cleanup.

Let's look at what happens when we create a goroutine:

```
func f() {
    fmt.Println("Hello from goroutine")
}

func main() {
    go f()
    fmt.Println("Hello from main")
    time.Sleep(100)
}
```

This program starts with the main goroutine. When the `go f()` statement is run, a new goroutine is created. Remember, a goroutine is an execution context, which means the `go` keyword causes the runtime to allocate a new stack and set it up to run the `f()` function. Then this goroutine is marked as ready to run. The main goroutine continues running without waiting for `f()` to be called, and prints `Hello from main` to console. Then it waits for 100 milliseconds. During this time, the new goroutine may start running, call `f()`, and print `Hello from goroutine`. `fmt.Println` has mutual exclusion built in to ensure that the two goroutines do not corrupt each other's outputs.

This program can output one of the following options:

- `Hello from main`, then `Hello from goroutine`: This is the case when the main goroutine first prints the output, then the goroutine prints it.

- `Hello from goroutine`, then `Hello from main`: This is the case when the goroutine created in `main()` runs first, and then the main goroutine prints the output.

- `Hello from main`: This is the case when the main goroutine continues running, but the new goroutine never finds a chance to run in the given 100 milliseconds, causing `main` to return. Once `main` returns, the program terminates without the goroutine ever finding a chance to run. It is unlikely that this case is observable, but it is possible.

Functions that take arguments can run as goroutines:

```go
func f(s string) {
    fmt.Printf("Goroutine %s\n", s)
}

func main() {
    for _, s := range []string{"a", "b", "c"} {
        go f(s)
    }
    time.Sleep(100)
}
```

Every run of this program is likely to print out a, b, and c in random order. This is because the `for` loop creates three goroutines, each called with the current value of s, and they can run in any order the scheduler picks them. Of course, if all goroutines do not finish within the given 100 milliseconds, some strings may be missing from the output.

Naturally, this can be done with an anonymous function as well. But now, things get interesting:

```go
func main() {
    for _, s := range []string{"a", "b", "c"} {
        go func() {
            fmt.Printf("Goroutine %s\n", s)
        }()
    }
    time.Sleep(100)
}
```

Here's the output:

```
Goroutine c
Goroutine c
Goroutine c
```

So, what is going on here?

First, this is a data race, because there is a shared variable that is written by one goroutine and read by three others without any synchronization. This becomes more evident if we unroll the `for` loop, as follows:

```go
func main() {
    var s string
    s = "a"
    go func() {
        fmt.Printf("Goroutine %s\n", s)
    }()

    s = "b"
    go func() {
        fmt.Printf("Goroutine %s\n", s)
    }()

    s = "c"
    go func() {
        fmt.Printf("Goroutine %s\n", s)
    }()

    time.Sleep(100)
}
```

In this example, each anonymous function is a closure. We are running three goroutines, each with a closure that captures the s variable from the enclosing scope. Because of that, we have three goroutines that read the shared s variable, and one goroutine (the main goroutine) writing to it concurrently. This is a data race. In the preceding run, all three goroutines ran after the last assignment to s. There are other possible runs. In fact, this program may even run correctly and print the expected output.

That is the danger of data races. A program such as this rarely runs correctly, so it is easy to diagnose and fix before code is deployed in a production environment. The data races that rarely give the wrong output usually make it to production and cause a lot of trouble.

Let's look at how closures work in more detail. They are the cause of many misunderstandings in Go development because simply refactoring a declared function as an anonymous function may have unexpected consequences.

A **closure** is a function with a context that includes some variables included in its enclosing scope. In the preceding example, there are three closures, and each closure captures the s variable from their scope. The **scope** defines all symbol names accessible at a given point in a program. In Go, the scope is determined syntactically, so where we declare the anonymous function, the scope includes all the exported functions, variables, type names, the main function, and the s variable. The Go compiler analyzes the source code to determine whether a variable defined in a function may be referenced after that function returns. This is the case when, for instance, you pass a pointer to a variable defined in one function to another function. Or when you assign a global pointer variable to a variable defined in a function. Once the function declaring that variable returns, the global variable will be pointing to a stale memory location. Stack locations come and go as functions enter and return. When such a situation is detected (or even a potential for such a situation is detected, such as creating a goroutine or calling another function), the variable escapes to the heap. That is, instead of allocating that variable on the stack, the compiler allocates the variable dynamically on the heap, so even if the variable leaves scope, its contents remain accessible. This is exactly what is happening in our example. The s variable escapes to the heap because there are goroutines that can continue running and accessing that variable even after main returns. This situation is depicted in *Figure 2.1*:

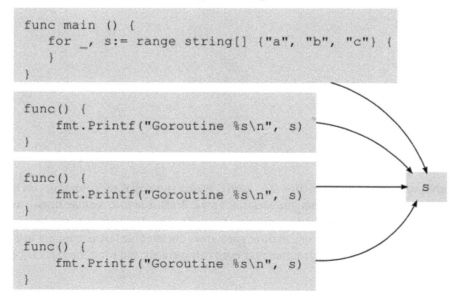

Figure 2.1 – Closures

Closures as goroutines can be a very powerful tool, but they must be used carefully. Most closures running as goroutines share memory, so they are prone to races.

We can fix our program by creating a copy of the s variable at each iteration. The first iteration sets s to "a". We create a copy of it and capture that copy in the closure. Then the next iteration sets s to "b". This is fine because the closure created during the first iteration is still using "a". We create a new copy of s, this time with a value of "b", and this goes on. This is shown in the following code:

```
for _, s := range []string{"a", "b", "c"} {
    s:=s // Redeclare s, create a copy
    // Here, the redeclared s shadows the loop variable s
    go func() {…}
}
```

Another way is to pass it as a parameter:

```
for _, s := range []string{"a", "b", "c"} {
    go func(s string) {
            fmt.Printf("Goroutine %s\n", s)
    }(s) // This will pass a copy of s to the function
}
```

In either solution, the s loop variable no longer escapes to the heap, because a copy of it is captured. In the first solution using a redeclared variable, the copy escapes to the heap, but the s loop variable doesn't.

One of the frequently asked questions regarding goroutines is: how do we stop a running goroutine? There is no magic function that will terminate or pause a goroutine. If you want to stop a goroutine, you have to send some message or set a flag shared with the goroutine, and the goroutine either has to respond to the message or read the shared variable and return. If you want to pause it, you have to use one of the synchronization mechanisms to block it. This fact causes some anxiety among developers who cannot find an effective way to terminate their goroutines. However, this is one of the realities of concurrent programming. The ability to create concurrent execution blocks is only one part of the problem. Once created, you have to be mindful of how to terminate them responsibly.

A panic can terminate a goroutine. If a panic happens in a goroutine, it is propagated up the call stack until a recover is found, or until the goroutine returns. This is called stack unwinding. If a panic is not handled, a panic message will be printed and the program will crash.

Before closing this topic, it might be helpful to talk about how Go runtime manages goroutines. Go uses an M:N scheduler that runs M goroutines on N OS threads. Internally, the Go runtime keeps track of the OS threads and the goroutines. When an OS thread is ready to execute a goroutine, the scheduler selects one that is ready to run and assigns it to the thread. The OS thread runs that goroutine until it blocks, yields, or is preempted. There are several ways a goroutine can be blocked. Blocking by channel operations or mutexes is managed by the Go runtime. If the goroutine is blocked because of a synchronous I/O operation, then the thread running that goroutine will also be blocked (this is managed by the operating system). In this case, the Go runtime starts a new thread or uses one already available and continues operation. When the OS thread unblocks (that is, the I/O operation ends), the thread is put back into use or returned to the thread pool. The Go runtime limits the number of active OS threads running user goroutines with the GOMAXPROCS variable. However, there is no limit on the number of OS threads waiting for I/O operations. So, the actual OS thread count a Go program uses can be much higher than GOMAXPROCS. However, only GOMAXPROCS of those threads would be executing user goroutines.

Figure 2.2 illustrates this. Suppose GOMAXPROCS=2. **Thread 1** and **Thread 2** are operating system threads that are executing goroutines. Goroutine **G1**, which is running on **Thread 1**, performs a synchronous I/O operation, blocking **Thread 1**. Since **Thread 1** is no longer operational, the Go runtime allocates **Thread 3** and continues running goroutines. Note that even though there are three operating system threads, there are two active threads and one blocked thread. When the system call running on **Thread 1** completes, the goroutine **G1** becomes runnable again, but there is one extra thread now. The Go runtime continues running with **Thread 3** and stops using **Thread 1**.

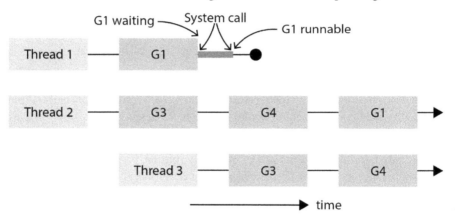

Figure 2.2 – System calls block OS threads

A similar process happens for asynchronous I/O operations, such as network operations and some file operations on certain platforms. However, instead of blocking a thread for a system call, the goroutine is blocked, and a netpoller thread is used to wait for asynchronous events. When the netpoller receives events, it wakes up the relevant goroutines.

Channels

Channels allow goroutines to share memory by communicating, as opposed to communicating by sharing memory. When you are working with channels, you have to keep in mind that channels are two things combined together: they are synchronization tools, and they are conduits for data.

You can declare a channel by specifying its type and its capacity:

```
ch:=make(chan int,2)
```

The preceding declaration creates and initializes a channel that can carry integer values with a capacity of 2. A channel is a **first-in, first-out (FIFO)** conduit. That is, if you send some values to a channel, the receiver will receive those values in the order they were written. Use the following syntax to send to or receive from channels:

```
ch <- 1    // Send 1 to the channel
<- ch      // Receive a value from the channel
x= <- ch   // Receive value from the channel and assign it to x
x:= <- ch  // Receive value from the channel, declare variable x
           // using the same type as the value read (which is
// int),and assign the value to x.
```

The len() and cap() functions work as expected for channels. The len() function will return the number of items waiting in the channel, and cap() will return the capacity of the channel buffer. The availability of these functions doesn't mean the following code is correct, though:

```
// Don't do this!
if len(ch) > 0 {
    x := <-ch
}
```

This code checks whether the channel has some data in it, and, seeing that it does, reads it. This code has a race condition. Even though the channel may have data in it when its length is checked, another goroutine may receive it by the time this goroutine attempts to do so. In other words, if len(ch) returns a non-zero value, it means that the channel had some values when its length was checked, but it doesn't mean that it has some values after the len function returns.

Figure 2.3 illustrates a possible sequence of operations with this channel using two goroutines. The first goroutine sends values **1** and **2** to the channel, which are stored in the channel buffer (len(ch)=2, cap(ch)=2). Then the other goroutine receives **1**. At this point, a value of 2 is the next one to be read from the channel, and the channel buffer only has one value in it. The first goroutine sends **3**. The channel is full, so the operation to send **4** to the channel blocks. When the second goroutine receives the 2 value from the channel, the send by the first goroutine succeeds, and the first goroutine wakes up.

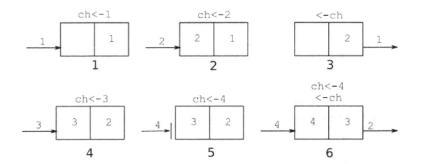

Figure 2.3 – Possible sequence of operations with a buffered channel of capacity 2

This example shows that *a send operation to a channel will block until the channel is ready to accept a value.* If the channel is not ready to accept the value, the send operation blocks.

Similarly, *Figure 2.4* shows a blocking receive operation. The first goroutine sends 1, and the second goroutine receives it. Now `len(ch)=0`, so the next receive operation by the second goroutine blocks. When the first goroutine sends a value of 2 to the channel, the second goroutine receives that value and wakes up.

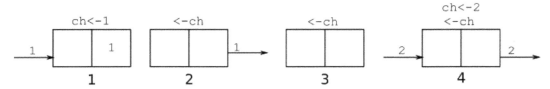

Figure 2.4 – Blocking receive operation

So, *a receive from a channel will block until the channel is ready to provide a value.*

A channel is actually a pointer to a data structure that contains its internal state, so the zero-value of a channel variable is `nil`. Because of that, channels must be initialized using the `make` keyword. If you forget to initialize a channel, it will never be ready to accept a value, or provide a value, thus *reading from or writing to a nil channel will block indefinitely.*

The Go garbage collector will collect channels that are no longer in use. If there are no goroutines that directly or indirectly reference a channel, the channel will be garbage collected even if its buffer has elements in it. You do not need to close channels to make them eligible for garbage collection. In fact, closing a channel has more significance than just cleaning up resources.

You may have noticed sending and receiving data using channels is a one-to-one operation: one goroutine sends, and another receives the data. It is not possible to send data that will be received by many goroutines using one channel. However, closing a channel is a one-time broadcast to all receiving goroutines. In fact, that is the only way to notify multiple goroutines at once. This is a very useful feature, especially when developing servers. For example, the `net/http` package implements a

`Server` type that handles each request in a separate goroutine. An instance of `context.Context` is passed to each request handler that contains a `Done()` channel. If, for example, the client closes the connection before the request handler can prepare the response, the handler can check to see whether the `Done()` channel is closed and terminate processing prematurely. If the request handler creates goroutines to prepare the response, it should pass the same context to these goroutines, and they will all receive the cancellation notice once the `Done()` channel closes. We will talk about how to use `context.Context` later in the book.

Receiving from a closed channel is a valid operation. In fact, *a receive from a closed channel will always succeed with the zero value of the channel type*. Writing to a closed channel is a bug: *writing to a closed channel will always panic*.

Figure 2.5 depicts how closing a channel works. This example starts with one goroutine sending 1 and 2 to the channel and then closing it. After the channel is closed, sending more data to it will cause a panic. The channel keeps the information that the channel is closed as one of the values in its buffer, so receive operations can still continue. The goroutine receives the 1 and 2 values, and then every read will return the zero value for the channel type, in this case, the 0 integer.

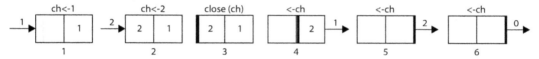

Figure 2.5 – Closing a channel

For a receiver, it is usually important to know whether the channel was closed when the read happened. Use the following form to test the channel state:

```
y, ok := <-ch
```

This form of channel receive operation will return the received value and whether or not the value was a real receive or whether the channel is closed. If `ok=true`, the value was received. If `ok=false`, the channel was closed, and the value is simply the zero value. A similar syntax does not exist for sending because sending to a closed channel will panic.

What happens when a channel is created without a buffer? Such a channel is called an **unbuffered channel**, and behaves in the same way as a buffered channel, but with `len(ch)=0` and `cap(ch)=0`. Thus, a send operation will block until another goroutine receives from it. A receive operation will block until another goroutine sends to it. In other words, an unbuffered channel is a way to transfer data between goroutines atomically. Let's look at how unbuffered channels are used to send messages and to synchronize goroutines using the following snippet:

```
1: chn := make(chan bool) // Create an unbuffered channel
2: go func() {
```

```
3:       chn <- true   // Send to channel
4: }()
5: go func() {
6:     var y bool
7:     y <-chn         // Receive from channel
8:     fmt.Println(y)
9: }()
```

Line 1 creates an unbuffered `bool` channel.

Line 2 creates the G1 goroutine and line 5 creates the G2 goroutine.

There are two possible runs at this point: **G1** attempts to send (line 3) before G2 is ready to receive (line 7), or G2 attempts to receive (line 7) before G1 is ready to send (line 3). The first diagram in *Figure 2.6* illustrates the case where **G1** runs first. At line 3, G1 attempts to send to the channel. However, at this point, **G2** is still not ready to receive. Since the channel is unbuffered and there are no receivers available, G1 blocks.

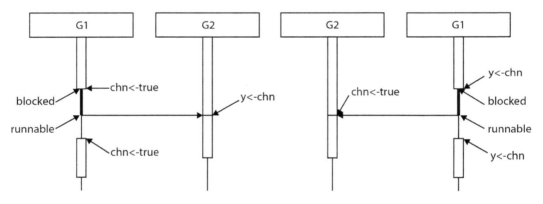

Figure 2.6 – Two possible runs using an unbuffered channel

After a while, G2 executes line 7. This is a channel receive operation, and there is a goroutine (G1) waiting to send to it. Because of this, the first G1 is unblocked and sends the value to the channel, and G2 receives it without blocking. It is now up to the scheduler to decide when G1 can run.

The second possible scenario, where G2 runs first, is shown in *Figure 2.6* on the right-hand side. Since G1 is not yet sent to the channel, G2 blocks. When G1 is ready to send, G2 is already waiting to receive, so G1 does not block and sends the value, and, G2 unblocks and receives the value. The scheduler decides when G2 can run again.

Note that an unbuffered channel acts as a synchronization point between two goroutines. Both goroutines must align for the message transfer to happen.

A word of caution is necessary here. Transferring a value from one goroutine to another transfers a copy of the value. So, if a goroutine runs ch<-x and sends the value of x, and another goroutine receives it with y<-ch, then this is equivalent to y=x, with additional synchronization guarantees. The crucial point here is that it *does not transfer the ownership of the value*. If the transferred value is a pointer, you end up with a shared memory system. Consider the following program:

```
type Data struct {
    Values map[string]interface{}
}


func processData(data Data,pipeline chan Data) {
    data.Values = getInitialValues()   // Initialize the
                                       // map
    pipeline <- data          // Send data to another
                              // goroutine for processing
    data.Values["status"] = "sent"     // Possible data
                                       // race!

}
```

The processData function initializes the Values map and then sends the data to another goroutine for processing. But a map is actually a pointer to a complex map structure. When data is sent through the channel, the receiver receives a copy of the pointer to the same map structure. If the receiving goroutine reads from or writes to the Values map, that operation will be concurrent with the write operation shown in the preceding code snippet. That is a data race.

So, as a convention, it is a good practice to assume that if a value is sent via a channel, the ownership of the value is also transferred, and you should not use a variable after sending it via a channel. You can redeclare it or throw it away. If you have to, include an additional mechanism, such as a mutex, so you can coordinate goroutines after the value becomes shared.

A channel can be declared with a direction. Such channels are useful as function arguments, or as function return values:

```
var receiveOnly <-chan int // Can receive, cannot
                           // write or close
var sendOnly chan<- int    // Can send, cannot read
                           // or close
```

The benefit of this declaration is type safety: a function that takes a send-only channel as an argument cannot receive from or close that channel. A function that gets a receive-only channel returned from a function can only receive data from that channel but cannot send data or close it, for example:

```
func streamResults() <-chan Data {
    resultCh := make(chan Data)
    go func() {
        defer close(resultCh)
        results := getResults()
        for _, result := range results {
            resultCh <- result
        }
    }()
    return resultCh
}
```

This is a typical way of streaming the results of a query to the caller. The function starts by declaring a bidirectional channel but returns it as a directional one. This tells the caller that it is only supposed to read from that channel. The streaming function will write to it and close it when everything is done.

So far, we have looked at channels in the context of two goroutines. But channels can be used to communicate with many goroutines. When multiple goroutines attempt to send to a channel or when multiple goroutines attempt to read from a channel, they are scheduled randomly. There are many implications of this simple rule.

You can create many worker goroutines, all receiving from a channel. Another goroutine sends work items to the channel, and each work item will be picked up by an available worker goroutine and processed. This is useful for worker pool patterns where many goroutines work on a list of tasks concurrently. Then, you can have one goroutine reading from a channel that is written by many worker goroutines. The reading goroutine will collect the results of computations performed by those goroutines. The following program illustrates this idea:

```
1: workCh := make(chan Work)
2: resultCh := make(chan Result)
3: done := make(chan bool)
4:
5: // Create 10 worker goroutines
6: for i := 0; i < 10; i++ {
7:     go func() {
8:         for {
```

```
 9:                // Get work from the work channel
10:                work := <- workCh
11:                // Compute result
12:                // Send the result via the result channel
13:                resultCh <- result
14:            }
15:      } ()
16: }
17: results := make([]Result, 0)
18: go func() {
19:      // Collect all the results.
20:      for _, i := 0; i < len(workQueue); i++ {
21:            results = append(results, <-resultCh)
22:      }
23:      // When all the results are collected, notify the done
channel
24:      done <- true
25: } ()
26: // Send all the work to the workers
27: for _, work := range workQueue {
28:      workCh <- work
29: }
30: // Wait until everything is done
31: <- done
```

This is an artificial example that illustrates how multiple channels can be used to coordinate work. There are two channels used for passing data around, workCh for sending work to goroutines, and resultCh to collect computed results. There is one channel, the done channel, to control program flow. This is required because we would like to wait until all the results are computed and stored in the slice before proceeding. The program starts by creating the worker goroutines and then creating a separate goroutine to collect the results. All these goroutines will be blocked, waiting to receive data (lines 10 and 21). The for loop at the main body will then iterate through the work queue and send the work items to the waiting worker goroutines (line 28). Each worker will receive the work (line 10), compute a result, and send the result to the collector goroutine (line 13), which will place them in a slice. The main goroutine will send all the work items and then block until it receives a value from the done channel (line 31), which will come after all the results are collected (line 24). As you can see, there is an ordering of channel operations in this program: $28 < 10 < 13 < 21 < 24 < 31$. These types of orderings will be crucial in analyzing the concurrent execution of programs.

You may have noticed that in this program, all the worker goroutines leak; that is, they were never stopped. A good way to stop them is to close the work channel once we're done writing to it. Then we can check whether the channel is closed in the worker:

```
for _, work := range workQueue {
    workCh <- work
}
close(workCh)
```

This will notify the workers that the work queue has been exhausted and the work channel is closed. We change the workers to check for this, as shown in the following code:

```
work, ok := <- workCh
if !ok {           // Is the channel closed?
    return         // Yes. Terminate
}
```

There is a more idiomatic way of doing this. You can range over a channel in a `for` loop, which will exit when the channel is closed:

```
go func() {
    for work := range workCh { // Receive until channel
                              //closes
        // Compute result
        // Send the result via the result channel
        resultCh <- result
    }
}()
```

With this change, all the running worker goroutines will terminate once the work channel is closed.

We will explore these patterns in greater detail later in the book. However, for now, these patterns bring up another question: how do we work with multiple channels? To answer this, we have to introduce the `select` statement. The following definition is from the Go language specification:

A select statement chooses which of a set of possible send or receive operations proceed.

The `select` statement looks like a switch-case statement:

```
select {
    case x := <-ch1:
    // Received x from ch1
```

```
    case y := <-ch2:
    // Received y from ch2
    case ch3 <- z:
    // Sent z to ch3
    default:
    // Optional default, if none of the other
    // operations can proceed
}
```

At a high level, the select statement chooses one of the send or receive operations that can proceed and then runs the block corresponding to the chosen operation. Note the past tense in the previous comments. The block for the reception of x from ch1 runs only after x is received from ch1. If there are multiple send or receive operations that can proceed, the select statement chooses one randomly. If there are none, the select statement chooses the default option. If a default option does not exist, the select statement blocks until one of the channel operations becomes available.

It follows from the preceding definitions that the following blocks indefinitely:

```
select {}
```

Using the default option in a select statement is useful for non-blocking sends and receives. The default option will only be chosen when all other options are not ready. The following is a non-blocking send operation:

```
select {
case ch<-x:
    sent = true
default:
}
```

The preceding select statement will test whether the ch channel is ready for sending data. If it is ready, the x value will be sent. If it is not, the execution will continue with the default option. Note that this only means that the ch channel was not ready for sending when it was tested. The moment the default option starts running, send to ch may become available.

Similarly, the following is a non-blocking receive:

```
select {
case x = <- ch:
    received = true
default:
}
```

One of the frequently asked questions about goroutines is how to stop them. As explained before, there is no magic function that will stop a goroutine in the middle of its operation. However, using a non-blocking receive and a channel to signal a stop request, you can terminate a long-running goroutine gracefully:

```go
 1: stopCh := make(chan struct{})
 2: requestCh := make(chan Request)
 3: resultCh := make(chan Result)
 4: go func() {
 5:     for { // Loop indefinitely
 6:         var req Request
 7:         select {
 8:         case req = <-requestCh:
 9:             // Received a request to process
10:         case <-stopCh:
11:             // Stop requested, cleanup and return
12:             cleanup()
13:             return
14:         }
15:         // Do some processing
16:         someLongProcessing(req)
17:         // Check if stop requested before another long task
18:         select {
19:         case <-stopCh:
20:             // Stop requested, cleanup and return
21:             cleanup()
22:             return
23:         default:
24:         }
25:         // Do more processing
26:         result := otherLongProcessing(req)
27:         select {
28:         // Wait until resultCh becomes sendable, or stop requested
29:         case resultCh <- result:
30:             // Result is sent
31:         case <-stopCh:
```

```
32:                    // Stop requested
33:                    cleanup()
34:                    return
35:                }
36:        }
37:    }()
```

The preceding function works with three channels, one to receive requests from requestCh, one to send results to resultCh, and one to notify the goroutine of a request to stop stopCh. To send a stop request, the main goroutine simply closes the stop channel, which broadcasts all worker goroutines a request to stop.

The select statement at line 7 blocks until one of the channels, the request channel or the stop channel, has data to receive. If it receives from the stop channel, the goroutine cleans up and returns. If a request is received, then the goroutine processes it. The select statement at line 18 is a non-blocking read from the stop channel. If during the processing, stop is requested, it is detected here, and the goroutine can clean up and return. Otherwise, the processing continues, and a result is computed. The select statement at line 27 checks whether the listening goroutine is ready to receive the result or whether stop is requested. If the listening goroutine is ready, the result is sent, and the loop restarts. If the listening goroutine is not ready but stop is requested, the goroutine cleans up and returns. This select is a blocking select, so it will wait until it can transmit the result or receive the stop request and return. Note that for the select statement at line 27, if both the result channel and the stop channel are enabled, the choice is random. The goroutine may send the result channel and continue with the loop even if stop is requested. The same situation applies to the select statement in line 7. If both the request channel and the stop channel are enabled, the select statement may choose to read the request instead of stopping.

This example brings up a good point: in a select statement, all enabled channels have the same likelihood of being chosen; that is, there is no channel priority. Under heavy load, the previous goroutine may process many requests even after a stop is requested. One way to deal with such a situation is to double-check the higher-priority channel:

```
select {
case req = <-requestCh:
    // Received a request to process
    // Check if also stop requested
    select {
    case <- stopCh:
        cleanup()
        return
    default:
```

```
        }
    case <-stopCh:
        // Stop requested, cleanup and return
        cleanup()
        return
    }
}
```

This will re-check the `stop` request after receiving it from the `request` channel and return if `stop` is requested.

Also, note that the preceding implementations will lose the received request if they are stopped. If that is not desired side effect, then the cleanup process should put the request back into a queue.

Channels can be used to gracefully terminate a program based on a signal. This is important in a containerized environment where the orchestration platform may terminate a running container using a signal. The following code snippet illustrates this scenario:

```
var term chan struct{}
func main() {
    term = make(chan struct{})
    sig := make(chan os.Signal, 1)
    go func() {
        <-sig
        close(term)
    }()
    signal.Notify(sig, syscall.SIGINT, syscall.SIGTERM)

    go func() {
        for {
            select {
            case term:
                return
            default:
            }
            // Do work
        }
    }()
    // ...
}
```

This program will handle the interrupt and termination signals coming from the operating system by closing a global term channel. All workers check for the term channel and return whether the program is terminating. This gives the application the opportunity to perform cleanup operations before the program terminates. The channel that listens to the signals must be buffered because the runtime uses a non-blocking write to send signal messages.

Finally, let's take a closer look at some of the interesting properties of select statements that may cause some misunderstandings. For example, the following is a valid select statement. When the channel becomes ready to receive, the select statement will choose one of the cases randomly:

```
select {
    case <-ch:
    case <-ch:
}
```

It is possible that the channel send or receive operation is not the first thing in a case block, for example:

```
func main() {
    var i int
    f := func() int {
        i++
        return i
    }
    ch1 := make(chan int)
    ch2 := make(chan int)
    select {
    case ch1 <- f():
    case ch2 <- f():
    default:
    }
    fmt.Println(i)
}
```

The preceding program uses a non-blocking send. There are no other goroutines, so the channel send operations cannot be chosen, but the f() function is still called for both cases. This program will print 2.

A more complicated select statement is as follows:

```
func main() {
    ch1 := make(chan int)
    ch2 := make(chan int)
```

```
    go func() {
        ch2 <- 1
    }()
    go func() {
        fmt.Println(<-ch1)
    }()
    select {
     case ch1 <- <-ch2:
            time.Sleep(time.Second)
    default:
    }
}
```

In this program, there is a goroutine that sends to the ch2 channel, and a goroutine that receives from ch1. Both channels are unbuffered, so both goroutines will block at the channel operation. But the select statement has a case that receives a value from ch2 and sends it to ch1. What exactly is going to happen? Will the select statement make its decision based on the readiness of ch1 or ch2?

The select statement will immediately evaluate the arguments to the channel send operation. That means <-ch2 will run, without looking at whether it is ready to receive or not. If ch2 is not ready to receive, the select statement will block until it becomes ready *even though there is a default case*. Once the message from ch2 is received, the select statement will make its choice: if ch1 is ready to send the value, it will send it. If not, the default case will be selected.

Mutex

Mutex is short for **mutual exclusion**. It is a synchronization mechanism to ensure that only one goroutine can enter a critical section while others are waiting.

A mutex is ready to be used when declared. Once declared, a mutex offers two basic operations: lock and unlock. A mutex can be locked only once, so if a goroutine locks a mutex, all other goroutines attempting to lock it will block until the mutex is unlocked. This ensures only one goroutine enters a critical section.

Typical uses of mutexes are as follows:

```
var m sync.Mutex
func f() {
    m.Lock()
    // Critical section
    m.Unlock()
```

```
    }
func g() {
    m.Lock()
    defer m.Unlock()
    // Critical section
}
```

To ensure mutual exclusion for a critical section, the *mutex must be a shared object*. That is, a mutex defined for a particularly critical section must be shared by all the goroutines to establish mutual exclusion.

We will illustrate the use of mutexes with a realistic example. A common problem that has been solved many times is the caching problem: certain operations, such as expensive computations, I/O operations, or working with databases, are slow, so it makes sense to cache the results once you obtain them. But by definition, a cache is shared among many goroutines, so it must be thread-safe. The following example is a cache implementation that loads objects from a database and puts them in a map. If the object does not exist in the database, the cache also remembers that:

```
type Cache struct {
    mu sync.Mutex
    m map[string]*Data
}

func (c *Cache) Get(ID string) (Data, bool) {
    c.mu.Lock()
    data, exists := c.m[ID]
    c.mu.Unlock()
    if exists {
        if data == nil {
            return Data{}, false
        }
        return *data, true
    }
    data, loaded = retrieveData(ID)
    c.mu.Lock()
    defer c.mu.Unlock()
    d, exists := c.m[data.ID]
    if exists {
        return *d, true
    }
```

```
    if !loaded {
        c.m[ID] = nil
        return Data{}, false
    }
    c.m[data.ID] = data
    return *data, true
}
```

The `Cache` structure includes a mutex. The `Get` method starts with locking the cache. This is because `Cache.m` is shared between goroutines, and all read or write operations involving `Cache.m` must be done by only one goroutine. If there are other cache requests ongoing at that moment, this call will block until the other goroutines are done.

The first critical section simply reads the map to see whether the requested object is already in the cache. Note the cache is unlocked as soon as the critical section is completed to allow other goroutines to enter their critical sections. If the requested object is in the cache, or if the nonexistence of that object is recorded in the cache, the method returns. Otherwise, the method retrieves the object from the database. Since the lock is not held during this operation, other goroutines may continue using the cache. This may cause other goroutines to load the same object as well. Once the object is loaded, the cache is locked again because the loaded object must be put in the cache. This time, we can use `defer c.mu.Unlock()` to ensure the cache is unlocked once the method returns. There is a second check to see whether the object was already placed in the cache by another goroutine. This is possible because multiple goroutines can ask for the object using the same ID at the same time, and many goroutines may proceed to load the object from the database. Checking this again after acquiring the lock will make sure that if another goroutine has already put the object into the cache, it will not be overwritten with a new copy.

An important point to note here is that mutexes should not be copied. When you copy a mutex, you end up with two mutexes, the original and the copy, and locking the original will not prevent the copies from locking their copies as well. The `go vet` tool catches these. For instance, declaring the cache `Get` method using a value receiver instead of a pointer will copy the cache struct and the mutex:

```
func (c Cache) Get(ID string) (Data,bool) {...}
```

This will copy the mutex at every call, thus all concurrent `Get` calls will enter into the critical section with no mutual exclusion.

A mutex does not keep track of which goroutine locked it. This has some implications. First, locking a mutex twice from the same goroutine will deadlock that goroutine. This is a common problem with multiple functions that can call each other and also lock the same mutex:

```
var m sync.Mutex
func f() {
```

```
        m.Lock()
        defer m.Unlock()
        // process
    }

    func g() {
        m.Lock()
        defer m.Unlock()
        f() // Deadlock
    }
```

Here, the g() function calls the f() function, but the m mutex is already locked, so f deadlocks. One way to correct this problem is to declare two versions of f, one with a lock and one without:

```
    func f() {
        m.Lock()
        defer m.Unlock()
        fUnlocked()
    }

    func fUnlocked() {
        // process
    }

    func g() {
        m.Lock()
        defer m.Unlock()
        fUnlocked()
    }
```

Second, there is nothing preventing an unrelated goroutine from unlocking a mutex locked by another goroutine. Such things tend to happen after refactoring algorithms and forgetting to change the mutex names during the process. They create very subtle bugs.

The functionality of a mutex can be replicated using a channel with a buffer size of 1:

```
    var mutexCh = make(chan struct{},1)
    func Lock() {
        mutexCh<-struct{}{}
```

```
}

func Unlock() {
    select {
    case <-mutexCh:
    default:
    }
}
```

Many times, such as in the preceding cache example, there are two types of critical sections: one for the readers and one for the writers. The critical section for the readers allows multiple readers to enter the critical section but does not allow a writer to go into the critical section until all readers are done. The critical section for writers excludes all other writers and all readers. This means that there can be many concurrent readers of a structure, but there can be only one writer. For this, an RWMutex mutex can be used. This mutex allows multiple readers or a single writer to hold the lock. The modified cache is shown as follows:

```
type Cache struct {
    mu sync.RWMutex // Use read/write mutex
    cache map[string]*Data
}

func (c *Cache) Get(ID string) (Data, bool) {
c.mu.RLock()
    data, exists := c.m[data.ID]
    c.mu.RUnlock()
    if exists {
        if data == nil {
            return Data{}, false
        }
    return *data, true
    }
    data, loaded = retrieveData(ID)
    c.mu.Lock()
    defer c.mu.Unlock()
    d, exists := c.m[data.ID]
    if exists {
        return *d, true
```

```
    }
    if !loaded {
        c.m[ID] = nil
                return Data{}, false
    }
    c.m[data.ID] = data
    return *data, true
}
```

Note that the first lock is a reader lock. It allows many reader goroutines to execute concurrently. Once it is determined that the cache needs to be updated, a writer lock is used.

Wait groups

A wait group waits for a collection of things, usually goroutines, to finish. It is essentially a thread-safe counter that allows you to wait until the counter reaches zero. A common pattern for their usage is this:

```
// Create a waitgroup
wg := sync.WaitGroup{}
for i := 0; i < 10; i++ {
    // Add to the wait group **before** creating the
    //goroutine
    wg.Add(1)
    go func() {
        // Make sure the waitgroup knows about
        // goroutine completion
         defer wg.Done()
        // Do work
    }()
}
// Wait until all goroutines are done
wg.Wait()
```

When you create a `WaitGroup`, it is initialized to zero, so a call to `Wait` will not wait for anything. So, you have to add the number of things it has to wait for before calling `Wait`. To do this, we call `Add(n)`, where n is the number of things to add for waiting. It makes it easier for the reader to call `Add(1)` just before creating the thing to wait, which is, in this case, a goroutine. The main goroutine then calls `Wait`, which will wait until the wait group counter reaches zero. For that to happen, we have to make sure that the `Done` method is called for each goroutine that returns. Using a `defer` statement is the easiest way to ensure that.

A common use for a `WaitGroup` is in an orchestrator service that calls multiple services and collects the results. The orchestrator service has to wait for all the services to return to continue computation.

Look at the following example:

```
func orchestratorService() (Result1, Result2) {
    wg := sync.WaitGroup{}  // Create a WaitGroup
    wg.Add(1)      // Add the first goroutine
    var result1 Result1
    go func() {
        defer wg.Done() // Make sure waitgroup
                        // knows completion
        result1 = callService1() // Call service1
    }()
    wg.Add(1)      // Add the second goroutine
    var result2 Result2
    go func() {
        defer wg.Done()  // Make sure waitgroup
                         // knows completion
        result2 = callService2()  // Call service2
    }()
    wg.Wait()     // Wait for both services to return
    return result1, result2    // Return results
}
```

A common mistake when working with a `WaitGroup` is calling Add or Done at the wrong place. There are two points to keep in mind:

- Add must be called before the program has a chance to run `Wait`. That implies that you cannot call Add inside the goroutine you are waiting for using the `WaitGroup`. There is no guarantee that the goroutine will run before `Wait` is called.

- Done must be called eventually. The safest way to do it is to use a `defer` statement inside the goroutine, so if the goroutine logic changes in time or it returns in an unexpected way (such as a `panic`), Done is called.

Sometimes using a wait group and channels together can cause some chicken or the egg problems: you have to close a channel after `Wait`, but `Wait` will not terminate unless you close the channel. Look at the following program:

```
1: func main() {
2:     ch := make(chan int)
```

```
 3:      var wg sync.WaitGroup
 4:      for i := 0; i < 10; i++ {
 5:          wg.Add(1)
 6:          go func(i int) {
 7:              defer wg.Done()
 8:              ch <- i
 9:          }(i)
10:      }
11:      // There is no goroutine reading from ch
12:      // None of the goroutines will return
13:      // so this will deadlock at Wait below
14:      wg.Wait()
15:      close(ch)
16:      for i := range ch {
17:          fmt.Println(i)
18:      }
19: }
```

One possible solution is to put the `for` loop at lines 16-18 into a separate goroutine before `Wait`, so there will be a goroutine reading from the channels. Since the channels will be read, all goroutines will terminate, which will release the `wg.Wait`, and close the channel, terminating the reader `for` loop:

```
go func() {
    for i := range ch {
        fmt.Println(i)
    }
}()
wg.Wait()
close(ch)
```

Another solution is as follows:

```
go func() {
    wg.Wait()
    close(ch)
}()
for i := range ch {
    fmt.Println(i)
}
```

The wait group is now waiting inside another goroutine, and after all the waited-for goroutines return, it closes the channel.

Condition variables

Condition variables differ from the previous concurrency primitives in the sense that, for Go, they are not an essential concurrency tool because, in most cases, a condition variable can be replaced with a channel. However, especially for shared memory systems, condition variables are important tools for synchronization. For example, the Java language builds one of its core synchronization features using condition variables.

A well-known problem of concurrent computing is the producer-consumer problem. There are one or more producer threads that produce a value. These values are consumed by one or more consumer threads. Since all producers and consumers are running concurrently, sometimes there are not enough values produced to satisfy all the consumers, and sometimes there are not enough consumers to consume values produced by the producers. There is usually a finite queue of values into which producers put, and from which consumers retrieve. There is already an elegant solution to this problem: use a channel. All producers write to the channel, and all consumers read from it, and the problem is solved. But in a shared memory system, a condition variable is usually employed for such a situation. A **condition variable** is a synchronization mechanism where multiple goroutines wait for a condition to occur, and another goroutine announces the occurrence of the condition to the waiting goroutines.

A condition variable supports three operations, as follows:

- `Wait`: Blocks the current goroutine until a condition happens
- `Signal`: Wakes up one of the waiting goroutines when the condition happens
- `Broadcast`: Wakes up all of the waiting goroutines when the condition happens

Unlike the other concurrency primitives, a condition variable needs a mutex. The mutex is used to lock the critical sections in the goroutines that modify the condition. It does not matter what the condition is; what matters is that the condition can only be modified in a critical section and that critical section must be entered by locking the mutex used to construct the condition variable, as shown in the following code:

```
lock := sync.Mutex{}
cond := sync.NewCond(&lock)
```

Now let's implement the producers-consumers problem using this condition variable. Our producers will produce integers and place them in a circular queue. The queue has a finite capacity, so the producer must wait until a consumer consumes from the queue if the queue is full. That means we need a condition variable that will cause the producers to wait until a consumer consumes a value. When the consumer consumes a value, the queue will have more space, and the producer can use it, but then the consumer who consumed that value has to signal the waiting producers that there is space available. Similarly, if consumers consume all the values before producers can produce new ones, the consumers have to wait until new values are available. So, we need another condition variable that will cause the consumers to wait until a producer produces a value. When a producer produces a new value, it has to signal to the waiting consumers that a new value is available.

Let's start with a simple circular queue implementation:

```go
type Queue struct {
    elements      []int
    front, rear int
    len         int
}


// NewQueue initializes an empty circular queue
//with the given capacity
func NewQueue(capacity int) *Queue {
    return &Queue{
    elements: make([]int, capacity),
    front:    0,  // Read from elements[front]
    rear:     -1, // Write to elements[rear]
    len:      0,
    }
}


// Enqueue adds a value to the queue. Returns false
// if queue is full
func (q *Queue) Enqueue(value int) bool {
    if q.len == len(q.elements) {
        return false
    }
    // Advance the write pointer, go around in a circle
    q.rear = (q.rear + 1) % len(q.elements)
    // Write the value
    q.elements[q.rear] = value
    q.len++
    return true
```

```
    }

    // Dequeue removes a value from the queue. Returns 0,false
    // if queue is empty
    func (q *Queue) Dequeue() (int, bool) {
        if q.len == 0 {
            return 0, false
        }
        // Read the value at the read pointer
        data := q.elements[q.front]
        // Advance the read pointer, go around in a circle
        q.front = (q.front + 1) % len(q.elements)
        q.len--
        return data, true
    }
```

We need a lock, two condition variables, and a circular queue:

```
func main() {
    lock := sync.Mutex{}
    fullCond := sync.NewCond(&lock)
    emptyCond := sync.NewCond(&lock)
    queue := NewQueue(10)
```

Here is the producer function. It runs in an infinite loop, producing random integer values:

```
producer := func() {
    for {
        // Produce value
        value := rand.Int()
        lock.Lock()
        for !queue.Enqueue(value) {
            fmt.Println("Queue is full")
            fullCond.Wait()
        }
        lock.Unlock()
        emptyCond.Signal()
        time.Sleep(time.Millisecond *
```

```
                time.Duration(rand.Intn(1000)))
    }
}
```

The producer generates a random integer, enters into its critical section, and attempts to enqueue the value. If it is successful, it unlocks the mutex and signals one of the consumers, letting it know that a value has been generated. If there are no consumers waiting on the emptyCond variable at that point, the signal is lost. If, however, the queue is full, then the producer starts waiting on the fullCond variable. Note that Wait is called in the critical section, with the mutex locked. When called, Wait atomically unlocks the mutex and suspends the execution of the goroutine. While waiting, the producer is no longer in its critical section, allowing the consumers to go into their own critical sections. When a consumer consumes a value, it will signal fullCond, which will wake one of the waiting producers up. When the producer wakes up, it will lock the mutex again. Waking up and locking the mutex is not atomic, which means, when Wait returns, the condition that woke up the goroutine may no longer hold, so Wait must be called inside a loop to recheck the condition. When the condition is rechecked, the goroutine will be in its critical section again, so no race conditions are possible.

The consumer is as follows:

```
consumer := func() {
for {
        lock.Lock()
        var v int
        for {
                var ok bool
                if v, ok = queue.Dequeue(); !ok {
                        fmt.Println("Queue is empty")
                        emptyCond.Wait()
                        continue
                }
                break
    }
        lock.Unlock()
        fullCond.Signal()
        time.Sleep(time.Millisecond *
           time.Duration(rand.Intn(1000)))
        fmt.Println(v)
    }
}
```

Note the symmetry between the producer and the consumer. The consumer enters into its critical section and attempts to dequeue a value inside a `for` loop. If the queue has a value in it, it is read, the `for` loop terminates, and the mutex is unlocked. Then the goroutine notifies any potential producers that a value is read from the queue, so it is likely that the queue is not full. By the time the consumer exists in its critical section and signals the producer, it is possible that another producer produced values to fill up the queue. That's why the producer has to check the condition again when it wakes up. The same logic applies to the consumer: if the consumer cannot read a value, it starts waiting, and when it wakes up, it has to check whether the queue has elements in it to be consumed.

The rest of the program is as follows:

```
for i := 0; i < 10; i++ {
    go producer()
}
for i := 0; i < 10; i++ {
    go consumer()
}
select {} // Wait indefinitely
```

You can run this program with different numbers of producers and consumers, and see how it behaves. When there are more producers than consumers, you should see more messages on the queue being full, and when there are more consumers than producers, you should see more messages on the queue being empty.

Summary

In this chapter, we introduced goroutines and channels, the two concurrency primitives supported by the Go language, as well as some of the fundamental synchronization primitives in the Go library. These primitives will be used in the next chapter to solve some of the popular concurrency problems.

Questions

1. Can you implement a mutex using channels? How about an RWMutex mutex?

2. Most condition variables can be replaced by channels. How would you implement Broadcast using channels?

3
The Go Memory Model

The Go memory model specifies when memory write operations become visible to other goroutines and, more importantly, when these visibility guarantees do not exist. As developers, we can skip over the details of the memory model by following a few guidelines when developing concurrent programs. Regardless, as mentioned before, obvious bugs are caught easily in QA, and bugs that happen in production usually cannot be reproduced in the development environment, and you may be forced to analyze the program behavior by reading code. A good knowledge of the memory model is helpful when this happens.

In this chapter, we will discuss the following:

- Why a memory model is necessary
- The happened-before relationship between memory operations
- Synchronization characteristics of Go concurrency primitives

Why a memory model is necessary

In 1965, Gordon Moore observed that the number of transistors in dense integrated circuits double every year. Later, in 1975, this was adjusted to doubling every 2 years. Because of these advancements, it quickly became possible to squeeze lots of components into a tiny chip, enabling the building of faster processors.

Modern processors use many advanced techniques, such as caching, branch prediction, and pipelining, to utilize the circuitry on a CPU to its maximum potential. However, in the 2000s, hardware engineers started to hit the limit of what could be optimized on a single chip. As a result, they created chips containing multiple cores. Nowadays, most performance considerations are about how fast a single core can execute instructions, as well as how many cores can run those instructions simultaneously.

The compiler technology did not stand still while these improvements were happening. Modern compilers can aggressively optimize programs to such an extent that the compiled code is unrecognizable. In other words, the order and the manner in which a program is executed may be very different from the way its statements are written. These reorganizations do not affect the behavior of a sequential program, but they may have unintended consequences when multiple threads are involved.

As a general rule, optimizations must not change the behavior of valid programs – but how can we define what a valid program is? The answer is a memory model. A memory model defines what a valid program is, what the compiler builders must ensure, and what programmers can expect.

In other words, a memory model is the compiler builder's answer to the hardware builder. As developers, we have to understand this answer so that we can create valid programs that run the same way on different platforms.

The Go memory model document starts with a rule of thumb for valid programs: **programs that modify data being simultaneously accessed by multiple goroutines must serialize such access**. It is as simple as that – but why is this so hard? The main reason comes down to figuring out when the effects of program statements can be observed at runtime, and the limits of an individual's cognitive capability: you simply cannot analyze all the possible orderings of a concurrent program. Familiarity with the memory model helps pinpoint problems in concurrent programs based on its observed behavior.

The Go memory model famously includes this phrase:

If you must read the rest of this document [i.e. the Go memory model] to understand the behavior of your program, you are being too clever.

Don't be clever.

I read this phrase as do not write programs that rely on the intricacies of the Go memory model. You should read and understand the Go memory model. It tells you what to expect from the runtime, and what not to do. If more people read it, there would be fewer questions related to Go concurrency on StackOverflow.

The happened-before relationship between memory operations

It all comes down to how memory operations are ordered at runtime, and how the runtime guarantees when the effects of those memory operations are observable. To explain the Go memory model, we need to define three relationships that define different orderings of memory operations.

In any goroutine, the ordering of memory operations must correspond to the correct sequential execution of that goroutine as determined by the control flow statements and expression evaluation order. This ordering is the **sequenced-before** relationship. This, however, does not mean that the compiler has to execute a program in the order it is written. The compiler can rearrange the execution order of statements as long as *a memory read operation of a variable reads the last value written to that variable.*

Let's refer to the following program:

```
1: x=1
2: y=2
3: z=x
4: y++
5: w=y
```

The z variable will always be set to 1, and the w variable will always be set to 3. Lines 1 and 2 are memory write operations. Line 3 reads x and writes z. Line 4 reads y and writes y. Line 5 reads y and writes w. That is obvious. One thing that may not be obvious is that the compiler can rearrange this code in such a way that all memory read operations read the latest written value, as follows:

```
1: x=1
2: z=x
3: y=2
4: y++
5: w=y
```

In these programs, every line is sequenced before the line that comes after it. The sequenced-before relationship is about the runtime order of statements and takes into account the control flow of the program. For instance, see the following:

```
1: y=0
2: for i:=0;i<2;i++ {
3:     y++
4: }
```

In this program, if y is 1 at the beginning of the statement at line 3, then the i++ statement that ran when i=0 is sequenced before line 3.

These are all ordinary memory operations. There are also synchronizing memory operations, which are defined as follows:

- **Synchronizing read operations**: Mutex lock, channel receive, atomic read, and atomic compare-and-swap operations

- **Synchronizing write operations**: Mutex unlock, channel send and channel close, atomic write, and atomic compare-and-swap operations

Note that atomic compare-and-swap is both a synchronizing read and a synchronizing write.

Ordinary memory operations are used to define the sequenced-before relationship within a single goroutine. Synchronizing memory operations can be used to define the synchronized-before relationship when multiple goroutines are involved. That is: *if a synchronizing memory read operation of a variable observes the last synchronized write operation to that variable, then that synchronized write operation is synchronized before the synchronized read operation.*

The **happened-before** relationship is a combination of a synchronized-before and sequenced-before relationships as follows:

- If a memory write operation, *w*, is synchronized before a memory read operation, *r*, then *w* happened before *r*

- If a memory write operation, *x*, is sequenced before *w*, and a memory read operation, *y*, is sequenced after *r*, then *x* happened before *y*

This is illustrated in *Figure 3.1* for the following program:

```go
go func() {
    x = 1
    ch <- 1
}()
go func() {
    <-ch
    fmt.Println(x)
}()
```

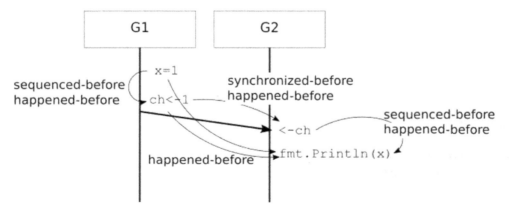

Figure 3.1 – Sequenced-before, synchronized-before, happened-before

The following modification, however, has a data race. After the first goroutine sends 1 to the channel and the second one receives it, x++ and fmt.Println(x) are concurrent:

```go
go func() {
    for {
        x ++
        ch <- 1
    }
}()
go func() {
    for range ch {
        fmt.Println(x)
    }
}()
```

The happened-before relationship is important because a memory-read operation is guaranteed to see the effects of memory-write operations that happened before it. If you suspect that a race condition exists, establishing which operations happened before others can help pinpoint any problems.

The rule of thumb is that if a memory-write operation happens before a memory-read operation, it cannot be concurrent, and thus, it cannot cause a data race. If you cannot establish a happened-before relationship between a write operation and a read operation, they are concurrent.

Synchronization characteristics of Go concurrency primitives

Having defined a happened-before relationship, it is easy to lay the ground rules for the Go memory model.

Package initialization

If package A imports another package, B, then the completion of all init() functions in package B happens before the init() functions in package A begin.

The following program always prints B initializing before A initializing:

```go
package B

import "fmt"

func init() {
```

```
    fmt.Println("B initializing")
}
---
package A

import (
    "fmt"
     "B"
)

func init() {
   fmt.Println("A initializing")
}
```

This extends to the main package as well: all packages directly or indirectly imported by the main package of a program complete their init() functions before main() starts. If an init() function creates a goroutine, there is no guarantee that the goroutine will finish before main() starts running. Those goroutines run concurrently.

Goroutines

If a program starts a goroutine, the go statement is synchronized before (and thus happens before) the start of the goroutine's execution. The following program will always print Before goroutine because the assignment to a happens before the goroutine starts running:

```
a := "Before goroutine"
go func() { fmt.Println(a) }()
select {}
```

The termination of a goroutine is not synchronized with any event in the program. The following program may print 0 or 1. There is a data race:

```
var x int
go func() { x = 1 }()
fmt.Println(x)
select {}
```

In other words, a goroutine sees all the updates before it starts, and a goroutine cannot say anything about the termination of another goroutine unless there is explicit communication with it.

Channels

A send or close operation on an unbuffered channel is synchronized before (and thus happens before) the completion of a receive from that channel. A receive operation on an unbuffered channel is synchronized before (and thus happens before) the completion of a corresponding send operation on that channel. In other words, if a goroutine sends a value through an unbuffered channel, the receiving goroutine will complete the reception of that value, and then the sending goroutine will finish sending that value. The following program always prints 1:

```
var x int
ch := make(chan int)
go func() {
    <-ch
    fmt.Println(x)
}()
x = 1
ch <- 0
select {}
```

In this program, the write to x is sequenced before the channel write, which is synchronized before the channel read. The printing is sequenced after the channel read, so the write to x happens before the printing of x.

The following program also prints 1:

```
var x int
ch := make(chan int)
go func() {
    ch <- 0
    fmt.Println(x)
}()
x = 1
<- ch
select {}
```

How can this guarantee be extended to buffered channels? If a channel has a capacity of C, then multiple goroutines can send C items before a single receive operation completes. In fact, the nth channel receive is synchronized before the completion of n + Cth send on that channel.

This is illustrated in *Figure 3.2* for a channel with a capacity of **2**.

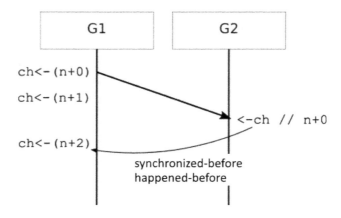

Figure 3.2 – Happened-before guarantees for buffered channels (capacity = 2)

This program creates 10 worker goroutines that share a resource with 5 instances. All 10 workers process jobs from the queue, but, at most, 5 of them can be working with the resource at any given time:

```
resources := make(chan struct{}, 5)
jobs := make(chan Work)
for worker := 0; worker < 10; worker++ {
    go func() {
        for work := range <-jobs {
            // Do work
            // Acquire resource
            resources <- struct{}{}
            // Work with resource
            <-resources
        }
    }()
}
```

In short, after a channel receive, the receiving goroutine sees all the updates before the corresponding send, and after a channel send, the sending goroutine sees all the operations before the corresponding receive.

Mutexes

Suppose two goroutines, G1 and G2, attempt to enter their critical sections by locking a shared mutex. Also, suppose that G1 locks the mutex (the first call to Lock()), and G2 is blocked. When G1 unlocks the mutex (the first call to Unlock()), G2 locks it (the second call to Lock()). Unlocking G1 is synchronized before (and thus, happens before) G2 is locked. As a result, G2 can see the effects of memory write operations G1 has made during its critical section.

In general, for a mutex, M, the nth call of M.Unlock() is synchronized before (and thus happens before) the (n + i)th M.Lock() returns when i > 0:

```go
func main() {
    var m sync.Mutex
    var a int
    // Lock mutex in main goroutine
    m.Lock()
    done := make(chan struct{})
    // G1
    go func() {
        // This will block until G2 unlocks mutex
        m.Lock()
        // a=1 happened-before, so this prints 1
        fmt.Println(a)
        m.Unlock()
        close(done)
    }()
    // G2
    go func() {
        a = 1
        // G1 will block until this runs
        m.Unlock()
    }()
    <-done
}
```

This program will always print 1, even if the first goroutine runs first.

To summarize: if a mutex lock succeeds, everything happened before the corresponding unlock is visible.

Atomic memory operations

The sync/atomic package provides low-level atomic memory read and memory write operations. If the effect of an atomic write operation is observed by an atomic read operation, then the atomic write operation is synchronized before that atomic read operation.

The following program always prints 1. This is because the print statement in the if block will only run after the atomic store operation completes:

```
func main() {
    var i int
    var v atomic.Value
    go func() {
            // This goroutine will eventually store 1 in v
        i = 1
        v.Store(1)
    }()
    go func() {
        // busy-waiting
        for {
            // This will keep checking until v has 1
            if val, _ := v.Load().(int); val == 1 {
                fmt.Println(i)
                return
            }
        }
    }()
    select {}
}
```

Map, Once, and WaitGroup

These are higher-level synchronization utilities that can be modeled by the operations explained previously. sync.Map provides a thread-safe map implementation that you can use without an additional mutex. This map implementation can outperform the built-in map coupled with a mutex if elements are written once but read many times, or if multiple goroutines work with disjoint sets of keys. For these use cases, sync.Map offers better concurrent use, but without the type safety of the built-in maps. Caches are good use cases for sync.Map, and an example of a simple cache is illustrated here.

For sync.Map, write operations happen before a read operation that observes the effects of that write.

sync.Once provides a convenient way of initializing something in the presence of multiple goroutines. The initialization is performed by a function passed to sync.Once.Do(). When multiple goroutines call sync.Once.Do(), one of the goroutines performs the initialization while the other goroutines block. After the initialization is complete, sync.Once.Do() no longer calls the initialization function, and it does not incur any significant overhead.

The Go memory model guarantees that if one of the goroutines results in the running of the initialization function, the completion of that function happens before the return of sync.Once() for all other goroutines.

The following is a cache implementation that uses sync.Map as a cache and uses sync.Once to ensure element initialization. Each cached data element contains a sync.Once instance, which is used to block other goroutines attempting to load the element with the same ID until initialization is complete:

```
type Cache struct {
    values sync.Map
}


type cachedValue struct {
    sync.Once
    value *Data
}


func (c *Cache) Get(id string) *Data {
    // Get the cached value, or store an empty value
    v, _:=c.values.LoadOrStore(id,&cachedValue{})
    cv := v.(*cachedValue)
    // If not initialized, initialize here
    cv.Do(func() {
        cv.value=loadData(id)
    })
    return cv.value
}
```

For `WaitGroup`, a call to `Done()` synchronizes before (and thus happens before) the return of the `Wait()` call that it unblocks. The following program always prints 1:

```
func main() {
    var i int
    wg := sync.WaitGroup{}
    wg.Add(1)
    go func() {
        i = 1
        wg.Done()
    }()
    wg.Wait()
    // If we are here, wg.Done is called, so i=1
    fmt.Println(i)
}
```

Let's summarize:

- Read operations of `sync.Map` return the last written value

- If many goroutines call `sync.Once`, only one will run the initialization and the other will wait, and once the initialization is complete, its effects will be visible to all waiting goroutines

- For `WaitGroup`, when `Wait` returns, all `Done` calls are completed

Summary

If you always serialize access to variables that are shared between multiple goroutines, you don't need to know all the details of the Go memory model. However, as you read, write, analyze, and profile concurrent code, the Go memory model can provide the insight that guides you to create safe and efficient programs.

Next, we will start working on some interesting concurrent algorithms.

Further reading

- *The Go Memory Model*: https://go.dev/ref/mem

- *Go's Memory Model*, a presentation by Russ Cox, February 25, 2016: http://nil.csail.mit.edu/6.824/2016/notes/gomem.pdf

4

Some Well-Known Concurrency Problems

This chapter is about some well-known concurrency problems that have many practical applications. The problems we will look at are as follows:

- The producer-consumer problem
- The dining philosophers problem
- Rate limiting

At the end of this chapter, you will have seen multiple implementations of these problems with some practical considerations on how to approach concurrency issues.

Technical Requirements

The source code for this particular chapter is available on GitHub at https://github.com/ PacktPublishing/Effective-Concurrency-in-Go/tree/main/chapter4.

The producer-consumer problem

In the previous chapter, we implemented a version of the producer-consumer problem using condition variables and mentioned that most of the time, the condition variables can be replaced by channels. The producer-consumer implementations we will work on in this chapter demonstrate this point. Some concurrency problems, such as the producer-consumer problem, are by their nature message-passing problems, and trying to solve them using shared-memory utilities results in unnecessarily complicated and lengthy code.

At the core of the producer-consumer problem is limited intermediate storage. At a high level, the producer-consumer problem contains processes that produce objects at various rates, and consumers that consume those objects at various rates, with limited storage in between the two that is used to store the produced objects until they are consumed. The producer-consumer problem is relevant in any system where a balance between the production of objects and their consumption must be established. For example, the goods produced at a factory must be stored somewhere until they are sold. If too much is produced, production must be slowed down. If there is too much demand, production must be increased.

We will start by reiterating the producer-consumer problem here. There are one or more producer goroutines that produce values, and there are one or more consumer goroutines that use these values in some way. We will write the producer goroutines so that they can be stopped using a signal from the main goroutine. When all the producers stop, the consumers should stop as well.

We will go through several iterations of the program. Our goal here is to illustrate how such a program can be implemented by starting from the simplest functional bits and then iteratively enhancing it to eventually develop a better version. So, the following is a good start for a producer:

```
func producer(index int, done <-chan struct{}, output chan<-
int) {
    for {
        // Produce a value
        value := rand.Int()
        // Wait a bit
        time.Sleep(time.Millisecond*
          time.Duration(rand.Intn(1000)))
        // Send the value
        select {
        case output <- value:
        case <-done:
          return
        }
        fmt.Printf("Producer %d sent %d\n", index, value)
    }
}
```

This function first generates a random value, waits a bit, and then sends the value to the channel. The index argument is simply used for printing which instance of the producer produced a particular value. When the value is sent to the channel, the function also checks if the done channel is triggered (by closing it), and if so, returns. This function will run in a goroutine, so returning from the function will also terminate the goroutine.

Now, let's write a `consumer` function:

```
func consumer(index int, input <-chan int) {
    for value := range input {
        fmt.Printf("Consumer %d received %d\n", index, value)
    }
}
```

The `consumer` function gets the consumer goroutine index and the data channel. It simply listens to the data channel and prints the values it receives. The `consumer` function will terminate when the input channel is closed.

Now, let's wire them up:

```
func main() {
    doneCh := make(chan struct{})
    dataCh := make(chan int, 0)
    for i := 0; i < 10; i++ {
        go producer(i, doneCh, dataCh)
    }
    for i := 0; i < 10; i++ {
        go consumer(i, dataCh)
    }
    select {}
}
```

This program creates a `data` channel and a `done` channel, starts 10 producers and 10 consumers, and runs indefinitely. You should notice the simplicity of this program as compared to the condition variable version. There is no concern for locking shared objects as there is none, nor is there a concern for buffering data produced by the producers. The channel takes care of all these issues. Multiple producers will put data on the channel, and the consumers will be randomly assigned to receive and process this data, all managed by the runtime.

But the program is not complete yet, because there is no way to terminate it gracefully. As a start, we can replace the `select{}` statement with a delay that will be used to run the program for a while (10 seconds) and then close the done channel:

```
// select {}
time.Sleep(time.Second * 10)
close(doneCh)
```

This is not enough, though. We closed the channel and broadcasted a request to terminate all the producers. Now, we have to wait for them to actually terminate. This can be done with `WaitGroup`:

```
producers := sync.WaitGroup{}
for i := 0; i < 10; i++ {
    producers.Add(1)
    go producer(i, &producers, doneCh, dataCh)
}
...
time.Sleep(time.Second * 10)
close(doneCh)
producers.Wait()
```

We have to change the producer function to accommodate this:

```
func producer(index int, wg *sync.WaitGroup, done chan
struct{}, output chan<- int) {
    defer wg.Done()
...
```

With these changes, we are now signaling the producers (`close(done)`) after running the program for 10 seconds, and then waiting for them to complete. Once they are complete, now we can signal the consumers to terminate. We do not use the `done` channel for this purpose, because we want the consumers to terminate only after they process all data elements. To do this, we will close `dataCh` once all producers are done. Closing `dataCh` will terminate the for-loops in the consumers, allowing them to return. This time, we have to wait for all of them to complete using a different wait group. The completed `main` function is here:

```
func main() {
    doneCh := make(chan struct{})
    dataCh := make(chan int)
    producers := sync.WaitGroup{}
    consumers := sync.WaitGroup{}
    for i := 0; i < 10; i++ {
        producers.Add(1)
        go producer(i, &producers, doneCh, dataCh)
    }
    for i := 0; i < 10; i++ {
        consumers.Add(1)
```

```
        go consumer(i, &consumers, dataCh)
    }
    time.Sleep(time.Second * 10)
    close(doneCh)
    producers.Wait()
    close(dataCh)
    consumers.Wait()
}
```

The obvious change for consumers is as follows:

```
func consumer(index int, wg *sync.WaitGroup, input <-chan int)
{
    defer wg.Done()
...
```

You might notice that using a simple channel reduces the complexity of the implementation considerably when compared to the version using condition variables.

Going back to the factory analogy at the beginning of this section, a channel quite accurately models "transferring goods between parties." By using channels with different capacities and adjusting the number of producers and consumers, you can fine-tune the behavior of the system for a particular load pattern. Keep in mind that such tuning and optimization activities should be performed only after you have a working implementation, and only after you measure the baseline behavior. Never attempt to optimize a program before observing how it runs. Make it run first, then you can make it good.

The dining philosophers problem

We visited the dining philosopher's problem in *Chapter 1, Concurrency: A High-Level Overview,* while discussing concurrency at a higher level. This is an important problem in the study of critical sections. The problem may seem contrived, but it shows a problem that comes up often in real-world situations: entering the critical section may require the acquisition of multiple resources (mutexes). Any time you have a critical section that relies on multiple mutexes, you have a chance of deadlock and starvation.

Now, we will study some solutions to this problem in Go. We will begin by restating the problem:

There are five philosophers dining together at the same round table. There are five plates, one in front of each philosopher, and one fork between each plate, five forks total. The dish they are eating requires them to use both forks, one on their left side and the other on their right side. Each philosopher thinks for a random interval and then eats for a while. To eat, a philosopher must acquire both forks, one on the left side and the other on the right side of the plate.

For our first solution, we use five goroutines representing the philosophers and five mutexes representing the forks. When a philosopher goroutine decides to eat, it must lock both mutexes. This model can be seen in *Figure 4.1*.

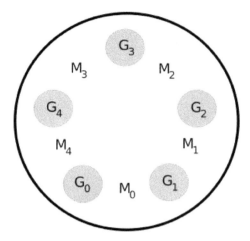

Figure 4.1 – The dining philosophers problem using goroutines and mutexes

The philosopher goroutine is as follows:

```
 1: func philosopher(index int, firstFork, secondFork *sync.
Mutex) {
 2:     for {
 3:             // Think for some time
 4:             fmt.Printf("Philosopher %d is thinking\n", index)
 5:             time.Sleep(time.Millisecond*time.Duration(rand.
Intn(1000)))
 6:             // Get the forks
 7:             firstFork.Lock()
 8:             secondFork.Lock()
 9:             // Eat
10:             fmt.Printf("Philosopher %d is eating\n", index)
11:             time.Sleep(time.Millisecond*time.Duration(rand.
Intn(1000)))
12:             secondFork.Unlock()
13:             firstFork.Unlock()
14:     }
15:}
```

In line 4, the philosopher thinks for a random amount of time. None of the forks have been picked up yet, so mutexes are not locked. Then, in line 7, the philosopher picks up the first fork. If the fork is already in use by the next philosopher, this philosopher blocks until the fork is released. Then, the philosopher picks up the second fork. Again, if the fork is being used by another philosopher, this philosopher must wait. After acquiring both forks, the philosopher eats for a random amount of time and releases both forks.

This implementation is prone to deadlock with the following `main` function:

```
func main() {
    forks := [5]sync.Mutex{}
    go philosopher(0, &forks[4], &forks[0])
    go philosopher(1, &forks[0], &forks[1])
    go philosopher(2, &forks[1], &forks[2])
    go philosopher(3, &forks[2], &forks[3])
    go philosopher(4, &forks[3], &forks[4])
    select {}
}
```

Analyzing an algorithm for deadlock comes down to finding where the goroutines can block. The philosopher goroutines can block at lines 7 and 8. Remember that for a deadlock to happen, one of the conditions is that at least one of the goroutines must hold a mutex exclusively (Coffman conditions). This means at least one of the goroutines must have successfully executed line 7 and locked a mutex. Another condition is that at least one goroutine must be waiting for a mutex held by another goroutine while it is holding one itself. That is, at least one goroutine must be blocked at line 8. This also implies that if there is a deadlock, at least one goroutine must be at line 8. The implementation guarantees the third Coffman condition: only the goroutine that locked a mutex unlocks it. Finding a deadlock then reduces to finding if a cyclic wait situation is reachable (the fourth Coffman condition).

Let's assume the system is deadlocked. We can tabulate which mutexes (forks) are held by each goroutine, and which mutexes the goroutine is blocked waiting on. *Figure 4.2* shows such a table. Here, G0 to G4 represent goroutines, and f0 to f4 represent the forks (mutexes) locked by that goroutine and the forks goroutine blocked waiting. We start by assuming a goroutine blocked attempting to lock a mutex and work our way backward to see if a deadlock state is reachable. For example, we start filling out the first row by entering G0 is blocked at f4. That means G4 locked f4 already. For this to happen, G4 must have also locked f3. That means G4 entered its critical section, and this cannot be a deadlock.

The second row shows a deadlock situation. G0 locked f4, but blocked on f0, G1 locked f0, but blocked on f1, and so on, and G4 locked f3 and blocked on f4. None of the goroutines can enter their critical sections because of this cyclic dependency. This is the fourth of the Coffman conditions, so a deadlock is actually possible. If all goroutines run line 7 one after the other before any one of them can run line 8, then the program deadlocks.

G0		G1		G2		G3		G4	
Locked	Blocked	Locked	Blocked	Locked	Blocked	Locked	Blocked	Locked	Blocked
	f4							f3, f4	
f4	f0	f0	f1	f1	f2	f2	f3	f3	f4

Figure 4.2 – Tabulating locked mutexes and blocked goroutines in search of a deadlock

Can we break the cycle? Indeed: if we can replace the order in which forks are picked up for one philosopher, the cycle breaks. For instance, if the first philosopher picks up the right fork first whereas all the others pick up the left fork, there is no deadlock:

```
func main() {
    forks := [5]sync.Mutex{}
    go philosopher(0, &forks[0], &forks[4])
    go philosopher(1, &forks[0], &forks[1])
    go philosopher(2, &forks[1], &forks[2])
    go philosopher(3, &forks[2], &forks[3])
    go philosopher(4, &forks[3], &forks[4])
    select {}
}
```

The goroutines are no longer identical though, so we have to do more work to show a deadlock cannot happen.

G0		G1		G2		G3		G4	
Locked	Blocked	Locked	Blocked	Locked	Blocked	Locked	Blocked	Locked	Blocked
	f0	f0	f1	f1	f2	f2	f3	f3, f4	
f0	f4							f3, f4	
f0			f0	f1	f2	f2	f3	f3, f4	
		f0		f0	f1				

Figure 4.3 – Tabulating locked mutexes and blocked goroutines in search of a deadlock

An exhaustive tabulation would require us to fill all possible options for every goroutine; that is, one option for blocking at line 7, and another option is for locking at line 7 and blocking at line 8. Because of the symmetrical implementation, we show the cases for G0 and G1 in *Figure 4.3*. In the first row, we assume G0 is blocked at line 7, which means G1 locked f0 and blocked at f1, which means G2 locked f1 and blocked at f2, and so on. In the end, we see that G4 can enter its critical section. Row 2 shows the case where G0 locked f0 but blocked at f4. Row 3 shows the case where G1 is blocked at locking f0 (line 7).

The fourth row shows an impossibility. We start this row by assuming G2 locked f0 but blocked on f1. This means G1 is blocked on f0. But that also means that G1 cannot lock f1, so G2 cannot be blocked on f1.

If you continue this process, you will observe that each row will either have an inconsistency like the 4th row, or one of the goroutines will be able to enter its critical section, which means there is no deadlock.

Many concurrency libraries, including later versions of Go, offer a `TryLock` method for mutexes. This might seem like an innocuous feature: if we cannot lock the mutex, do something else. In reality, there are surprisingly few cases where `TryLock` can be useful. You have to keep in mind that when you call `TryLock` and receive an indication that it could not lock the mutex, the mutex might be lockable. One possible use of `TryLock` is deadlock prevention:

```go
func philosopher(index int, leftFork, rightFork *sync.Mutex) {
    for {
        fmt.Printf("Philospher %d is thinking\n", index)
        time.Sleep(time.Millisecond*
                time.Duration(rand.Intn(1000)))
        // Get left fork
        leftFork.Lock()
        // Get right fork
        if rightFork.TryLock() {
            // Eat
            fmt.Printf("Philosopher %d is eating\n",
                    index)
            time.Sleep(time.Millisecond*
                    time.Duration(rand.Intn(1000)))
            rightFork.Unlock()
        }
        leftFork.Unlock()
    }
}
```

In this implementation, the philosopher goroutine picks up the left fork, then tries to pick up the right fork. If it fails, it puts the left fork back and continues thinking. This is deadlock-free, but prone to busy-spinning, causing starvation. It is possible that the philosopher goroutine spends a long time thinking, and picking up and then putting back the left fork. Every time the locking of the right fork fails, the left fork is released and locked, eliminating any advantage it might have had for waiting longer in the goroutine queue.

Could there be a deadlock-free implementation without using `TryLock` and without relying on the ordering of locks? For an answer, see the following channel implementation:

```go
func philosopher(index int, leftFork, rightFork chan bool) {
    for {
        // Think for some time
        fmt.Printf("Philospher %d is thinking\n", index)
        time.Sleep(time.Duration(rand.Intn(1000)))
        select {
        case <-leftFork:
            select {
            case <-rightFork:
                fmt.Printf("Philosopher %d
                        is eating\n", index)
                time.Sleep(time.Millisecond*
                    time.Duration(rand.Intn(1000)))
                rightFork <- true
            default:
            }
            leftFork <- true
        }
    }
}

func main() {
    var forks [5]chan bool
    for i := range forks {
        forks[i] = make(chan bool, 1)
        forks[i] <- true
    }
    go philosopher(0, forks[4], forks[0])
    go philosopher(1, forks[0], forks[1])
    go philosopher(2, forks[1], forks[2])
    go philosopher(3, forks[2], forks[3])
    go philosopher(4, forks[3], forks[4])
    select {}
}
```

In this implementation, each fork is modeled by a channel with a capacity of 1. When the fork is on the table, the channel has a value in it. That's why the program starts with initializing channels and placing the value into the channel (that is, putting the fork on the table). To take a fork off the table, the philosopher goroutine reads from the channel. To put the fork back, the philosopher goroutine writes to the channel. All philosophers wait until they take the left fork. Once they have the left fork, they try to take the right fork as well. If that is also successful, the philosopher eats. Otherwise, the left fork is put back on the table and the philosopher continues thinking. As you can see, this is almost identical to the mutex solution using `TryLock`, but with channels.

Rate limiting

Limiting the rate of requests for a resource is important to maintain a predictable quality of service. There are several ways rate control can be achieved. We will study two implementations of the same algorithm. The first one is a relatively simple implementation of the token bucket algorithm that uses channels, a ticker, and a goroutine. Then, we will study a more advanced implementation that requires fewer resources.

First, let's take a look at the token bucket algorithm and show how it is used for rate limiting. Imagine a fixed-sized bucket containing tokens. There is a producer process that deposits tokens into this bucket at a fixed rate, say two tokens/second. Every 500 milliseconds, this process adds a token to the bucket if the bucket has empty slots. If the bucket is full, it waits for another 500 milliseconds and checks the bucket again. There is also a consumer process that consumes tokens at random intervals. However, in order for the consumer process to continue, it has to take one token. If the bucket is empty, the consumer process has to wait for a token to be deposited. This is illustrated in *Figure 4.4*.

Figure 4.4 – The token bucket algorithm with limit=4, rate=2

To analyze how this structure can be used for rate limiting, first, we can look at how it behaves with a token bucket capacity of 1 and a rate of two tokens/second. Assuming requests come randomly and we start with a full bucket, the first request comes at time t=100ms and consumes the token. The next token will be delivered at time t=500ms, so any request that comes before that has to wait until the new token is produced. Suppose another request comes at t=300ms, and another at t=400ms. The first request can continue at t=500ms, when a token is generated, and the second request can continue at t=1000ms, when the next token is generated. Thus, no two requests can take place closer than 500ms apart, essentially limiting the rate of requests to two requests/second.

But what happens when the bucket size is larger than one? Say, the bucket can hold four tokens, as depicted in *Figure 4.4*. Again, starting with a full bucket, suppose we get a burst of requests at times t=100ms, t=110ms, t=120ms, t=130ms, and t=140ms. The bucket already contains four tokens, so the first four requests will consume those buckets and continue. But when it comes to the fifth request at time t=140ms, the bucket is empty, and that request has to wait until t=500 for a new token to arrive. Suppose the next requests come at t=1600ms, t=1700ms, and t=1800ms. The bucket has new tokens deposited to it at times t=1000ms and t=1500ms, so the first two requests can continue at t=1600ms and t=1700ms, but the next request has to wait for a new token until t=2000ms. So, a token bucket with a size larger than one allows bursts of that size to be processed, without violating the average rate requirement. For a given duration, the number of requests admitted is still the number of tokens deposited based on the rate. But the requests can arrive in bursts and the system admits that burstiness as long as the rate limit is not exceeded.

Based on what we have discussed, the rate limiter can implement an interface that looks like the following:

```
type RateLimit interface {
    Wait()
}
```

To rate limit an HTTP service, use a limiter shared between handlers. The `Wait` call will delay the handler until it is time to process the request:

```
func handle(w http.ResponseWriter,req *http.Request) {
    limiter.Wait()
    // Handle request
}
```

At this point, you may realize that the token bucket looks very much like a channel. Indeed, a channel offers a simple model to implement exactly what the algorithm describes. The channel becomes the token bucket, a producer goroutine uniformly puts tokens into the channel, and tokens can be consumed by reading from the channel. If the channel is empty, the `read` operation will block until a new token is deposited into the bucket. Therefore, we will need a channel and a ticker. We will also add another channel to close the limiter when it is done:

```
type ChannelRate struct {
    bucket   chan struct{}
    ticker *time.Ticker
    done     chan struct{}
}
```

We are using `struct{}` for the channel type here. `struct{}` occupies zero bytes, and Go handles this quite nicely, as expected, by not allocating any memory for the channel buffer. The `bucket` channel will hold the tokens and the done channel will only be used to shut down the rate limiter once we're done with it. We will need the ticker to generate periodic ticks so we can produce tokens. We start with a constructor function that initializes the structure members and fills the bucket:

```
 1:func NewChannelRate(rate float64, limit int) *ChannelRate {
 2:    ret := &ChannelRate{
 3:      bucket: make(chan struct{}, limit),
 4:      ticker: time.NewTicker(time.Duration(1 / rate *
1000000000)),
 5:      done:   make(chan struct{}),
 6:      }
 7:    for i := 0; i < limit; i++ {
 8:        ret.bucket <- struct{}{}
 9:      }
10:    go func() {
11:        for {
12:            select {
13:                case <-ret.done:
14:                    return
15:                case <-ret.ticker.C:
16:                    select {
17:                        case ret.bucket <- struct{}{}:
18:                        default:
19:                    }
20:                }
21:            }
22:    }()
23:    return ret
24:}
```

NewChannelRate gets two arguments. rate specifies the rate with which tokens will be generated in terms of rate tokens per second, and limit specifies the bucket size. So, we will initialize a channel with limit capacity and start a ticker with a 1/rate period (lines 2-6). Then, we will fill up the bucket to capacity (lines 7-9).

The remaining of this function is the goroutine that produces the tokens (lines 10-22). There are several points we need to be careful about here. First, if the bucket is full when we generate a token, the token should be thrown away (lines 16-19). Second, when the done channel is closed, we have to close the ticker and terminate the goroutine (lines 13-14).

Note the nonblocking send when a new token is generated. This ensures that the goroutine does not block if the bucket is full. Also, note that this goroutine is a closure, so it is connected to the rate limiter instance it is instantiated for.

The `Wait` method is now trivial to implement. The following method will wait until a token is available in the bucket:

```
func (s *ChannelRate) Wait() {
    <-s.bucket
}
```

We can gracefully close the ticker with the following method:

```
func (s *ChannelRate) Close() {
    close(s.done)
    s.ticker.Stop()
}
```

A potential problem with this limiter is that it requires two additional goroutines for every instance of the limiter: one for the producer, and one for the ticker. This is usually not a problem, especially if the rate limiter is being used to control access to a common service. It may start to get resource intensive if many instances of the rate limiter are required. For instance, many API providers use rate limiting based on the customer account. With this rate limiter, three goroutines are required for each request: one to do the actual work for each customer, and two more to rate limit it. Can this be done without creating any additional goroutines?

The answer is *yes*. The key to the solution is to realize that the actual rate limiting only happens when a token is consumed. So, instead of creating a goroutine to periodically fill the bucket, we can fill it only when we need it, that is, only when we want to consume a token and the bucket is empty. We can start by replacing the bucket channel with an integer value, nTokens, keeping the number of tokens in the bucket. Every time we consume a token, we decrease this number. The actual work is done when we attempt to consume a token but the bucket is empty; this is when `nTokens=0`.

First, consider the situation in *Figure 4.5*. The last token was generated at t_{last}. A request consumed that token, so the bucket is now empty. A new request comes at t_{req}, which is late enough that multiple tokens should have been generated between t_{last} and t_{req}. So, we simply calculate the number of tokens the bucket must have by $(t_{req}-t_{last})$/period. Then we consume one of those tokens and continue. The new value for the last token generated is t_{last}+nTokens*period.

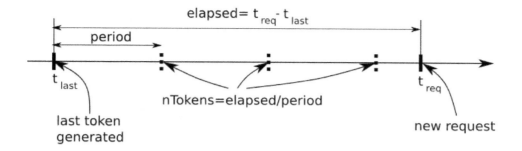

Figure 4.5 – A new request comes after multiple tokens are generated

Another possibility is shown in *Figure 4.6*. This is the situation when a new request comes before the next token can be generated. In this case, the request must wait until it is time to generate the new token, which is equal to $t_{last}+period-t_{req}$.

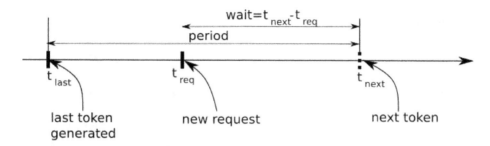

Figure 4.6 – New request comes before the next token is generated

Based on this, we have to change the limiter definition to the following:

```
type Limiter struct {
        mu sync.Mutex
        // Bucket is filled with rate tokens per second
        rate int
        // Bucket size
        bucketSize int
        // Number of tokens in bucket
        nTokens int
        // Time last token was generated
        lastToken time.Time
}
```

We now need a mutex to protect the rate limiter variables because we no longer have a channel that will ensure mutual exclusion. The `Wait` method can only be called by one goroutine and all others have to wait until the active goroutine is done. The initialization is simple:

```
func NewLimiter(rate, limit int) *Limiter {
    return &Limiter{
        rate:       rate,
        bucketSize: limit,
        nTokens:    limit,
        lastToken:  time.Now(),
    }
}
```

This time, we have to keep the rate and limit variables in the struct because we will use them to calculate when tokens are generated.

The `Wait` method is where everything happens:

```
 1: func (s *Limiter) Wait() {
 2:     s.mu.Lock()
 3:     defer s.mu.Unlock()
 4:     if s.nTokens > 0 {
 5:         s.nTokens--
 6:         return
 7:     }
 8:     // Here, there is not enough tokens in the bucket
 9:     tElapsed := time.Since(s.lastToken)
10:      period := time.Second / time.Duration(s.rate)
11:     nTokens := tElapsed.Nanoseconds() / period.Nanoseconds()
12:     s.nTokens = int(nTokens)
13:     if s.nTokens > s.bucketSize {
14:         s.nTokens = s.bucketSize
15:     }
16:     s.lastToken = s.lastToken.Add(time.Duration(nTokens) * period)
17:     // We filled the bucket. There may not be enough
18:     if s.nTokens > 0 {
19:         s.nTokens--
20:         return
```

```
21:     }
22:     // We have to wait until more tokens are available
23:     // A token should be available at:
24:     next := s.lastToken.Add(period)
25:     wait := next.Sub(time.Now())
26:     if wait >= 0 {
27:         time.Sleep(wait)
28:     }
29:     s.lastToken = next
30: }
```

Lines 4-7 are what is usually called the *happy path*. This is when there are tokens in the bucket, so we simply grab one and return. If the algorithm comes to line 9, then there are no tokens in the bucket. Lines 9-11 compute how many tokens should be generated based on the elapsed time since the last time a token was generated. If this number is larger than the bucket size, we can fill up the bucket and throw away the rest of the tokens (lines 13-15). Line 16 updates the last token generated time. Note that this uses the actual number of tokens generated, not just the ones that are stored in the bucket. At this point, if there is a token in the bucket, we grab it return (lines 18-21). If not, we have to wait. The amount of time to wait is calculated based on the current time and the next token generation time (lines 24-25). Then, the rate limiter waits and when the wait is over, we know there is only one token in the bucket, so it consumes the token and returns.

This rate limiter works without creating any additional goroutines. It is less resource intensive than the previous one, so more suitable to be used in a setting where many rate limiters are required, such as in an API provider where rate limiting is configured based on the users of the APIs. There is a publicly available rate limiter package, `golang.org/x/time/rate`, that uses a similar implementation to this one. For production use cases, use that package as it provides a much richer API and context support. Context support is necessary because, as you can see, our rate limiter continues waiting even if the requestor cancels the request.

Summary

In this chapter, we studied three well-known concurrency problems that show up consistently when working with non-trivial problems. Producer-consumer implementations have uses in data processing pipelines, crawlers, device interactions, network communications, and more. The dining philosophers problem is a good demonstration of critical sections that require multiple mutexes. Finally, rate-limiting has applications in ensuring the quality of service, limiting resource utilization, and API accounting.

In the next chapter, we will start looking at more realistic examples of concurrent programming, in particular, worker pools, concurrent data pipelines, and fan-in/fan-out.

5
Worker Pools and Pipelines

This chapter is about two interrelated concurrency constructs: worker pools and pipelines. While a worker pool deals with splitting work among multiple instances of the same computation, a data pipeline deals with splitting work into a sequence of different computations, one after the other.

In this chapter, you will see several working examples of worker pools and data pipelines. These patterns naturally come up as solutions to many problems, and there is no single best solution. I try to separate the concurrency concerns from the computation logic. If you can do the same for your problems, you can iteratively find the best solution for your use case.

The topics that this chapter will cover are as follows:

- Worker pools, using a file scanner example
- Data pipelines, using a CSV file processor example

Technical Requirements

The source code for this particular chapter is available on GitHub at `https://github.com/PacktPublishing/Effective-Concurrency-in-Go/tree/main/chapter5`.

Worker pools

Many concurrent Go programs are combinations of variations on worker pools. One reason could be that channels provide a really good mechanism for assigning tasks to waiting goroutines. A worker pool is simply a group of one or more goroutines that performs the same task on multiple instances of inputs. There are several reasons why a worker pool may be more practical than creating goroutines as needed. One reason is that creation of a worker instance in the worker pool could be expensive (not the creation of a goroutine, that's cheap, but the initialization of a worker goroutine can be expensive), so a fixed number of workers can be created once and then reused. Another reason is that you potentially need an unbounded number of them, so you create a reasonable number once. Regardless of the situation, once you decide you need a worker pool, there are easy-to-repeat patterns that you can use over and over to create high-performing worker pools.

We first saw a simple worker pool implementation in *Chapter 2*. Let's look at some variations of the same pattern here. We will work on a program that recursively scans directories and searches for regular expression matches, a simple **grep** utility. The Work struct is defined as a structure containing the filename and the regular expression:

```
type Work struct {
    file     string
    pattern *regexp.Regexp
}
```

In most systems, the number of files you can open is limited, so we work with a fixed-size worker pool:

```
func main() {
    jobs := make(chan Work)
    wg := sync.WaitGroup{}
    for i := 0; i < 3; i++ {
        wg.Add(1)
        go func() {
            defer wg.Done()
            worker(jobs)
        }()
    }
...
```

Note that we created a WaitGroup here, so we can wait for all workers to finish processing before the program exits. Also note that by using an anonymous function that wraps the actual worker, we can isolate the worker itself from the mechanics of the WaitGroup. Then we compile the regular expression that will be used by all goroutines:

```
rex, err := regexp.Compile(os.Args[2])
if err != nil {
    panic(err)
}
```

The rest of the main function walks the directories and sends files to the workers:

```
filepath.Walk(os.Args[1], func(path string, d fs.FileInfo, err
error) error {
    if err != nil {
        return err
```

```
        }
        if !d.IsDir() {
                jobs <- Work{file: path, pattern: rex}
        }
        return nil
})
```

And finally, we terminate all the workers and wait for them to complete:

```
    ...
    close(jobs)
    wg.Wait()
}
```

The actual `worker` function reads the file line by line and checks whether there are any pattern matches. If so, it prints the filename and the matching line:

```
func worker(jobs chan Work) {
    for work := range jobs {
            f, err := os.Open(work.file)
            if err != nil {
                    fmt.Println(err)
                    continue
            }
            scn := bufio.NewScanner(f)
            lineNumber := 1
            for scn.Scan() {
                    result := work.pattern.Find(scn.Bytes())
                    if len(result) > 0 {
                            fmt.Printf("%s#%d: %s\n", work.file,
                            lineNumber, string(result))
                    }
                    lineNumber++
            }
            f.Close()
    }
}
```

Note that the `worker` function continues until the `jobs` channel closes. When the program is run, each file is sent to the `worker` function, and the `worker` function processes the file. Since the worker pool has three workers in it, at any given moment, there will be at most three files that are being processed concurrently. Also note that the workers print the results concurrently, so the matching lines for each file are randomly interleaved.

This worker pool prints the results instead of returning them to the caller. In many cases, it is necessary to get the results from the worker pool after the work is submitted. A good way to do this is to include a return channel in the `Work` struct itself:

```go
type Work struct {
    file    string
    pattern *regexp.Regexp
    result  chan Result
}
```

We change the `worker` function to send the results via the result channel. Also, do not forget to close that result channel once the processing of the file is done, so the receiving end knows there will be no more results coming from that channel:

```go
. . .
for scn.Scan() {
    result := work.pattern.Find(scn.Bytes())
    if len(result) > 0 {
        work.result <- Result{
            file:       work.file,
            lineNumber: lineNumber,
            text:       string(result),
        }
    }
    lineNumber++
}
close(work.result)
```

This design solves the problem of interleaved results. We can read from one result channel until it is done, then move on to the next. But we cannot use the same goroutine to submit the jobs and read the results because that would cause a deadlock. Can you see why? We would be submitting jobs to a worker pool with no one listening for the results, so after we submit enough jobs to assign to every worker, the channel send operation will block. So, the goroutine that receives the results must be different from the one that sends them. I choose to put the directory walker in its own goroutine and read the results in the main goroutine.

There is one more problem to solve: how are we going to let the receiver goroutine know about the result channels? Every submitted work contains a new channel from which we have to read from. We can use a slice and add all those channels to it, but that slice will need synchronization because it will be read and written from multiple goroutines.

We can use a channel to send those results channels:

```
allResults := make(chan chan Result)
```

We will send every new result channel to the allResults channel. When the main goroutine receives that channel, it will iterate over it to print the results, and stop the iteration once the worker goroutine closes the result channel. Then, it will receive the next channel from allResults, and continue printing. The file walker now looks like this:

```
go func() {
    defer close(allResults)
    filepath.Walk(os.Args[1], func(path string,
      d fs.FileInfo, err error) error {
        if err != nil {
            return err
        }
        if !d.IsDir() {
            ch := make(chan Result)
            jobs <- Work{file: path, pattern: rex,
              result: ch}
            allResults <- ch
        }
        return nil
    })
}()
```

Note the defer statement at the beginning. Once all the files are sent, we close the allResults channel to signal the completion of processing. We read the results using the following code:

```
for resultCh := range allResults {
    for result := range resultCh {
        fmt.Printf("%s #%d: %s\n", result.file,
          result.lineNumber, result.text)
    }
}
```

Figure 5.1 shows how we can analyze this algorithm. We have three goroutines here, from left to right, the path walker, the worker, and the main goroutine. This figure only shows the synchronization points of these goroutines. Initially, the path walker starts running, finds a file, and attempts to send the `Work` struct to the `jobs` channel. The worker goroutine waits to receive from the `jobs` channel, and the main goroutine waits to receive from the `allResults` channel:

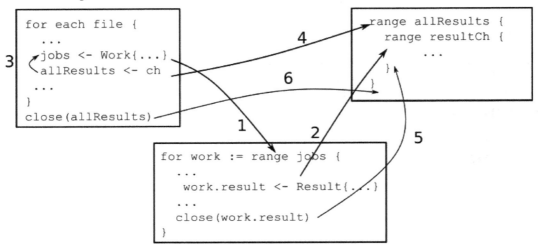

Figure 5.1 – Happened-before relationships for the worker pool

Now let's say there is a worker available, so the send to the `jobs` channel succeeds, and the worker receives the work (arrow 1). At this point, the path walker goes on to send to the `allResults` channel (arrow 3), which depends on the main goroutine to receive from that channel (arrow 4), so the path walker continues running, and the main goroutine starts waiting to receive results from `resultCh`. While all these are happening, the worker goroutine computes a result and writes it to the result channel for the work, which is then received by the main goroutine (arrow 2). This continues until the worker is done, so it closes the results channel, which terminates the loop in the main goroutine (arrow 5). Now the path walker is ready to send the next piece of work. When the path walker is done, it closes the `allResults` channel, which terminates the forloop in the main goroutine (arrow 6).

It is also possible to use a worker pool to perform computations whose results will be used later (similar to `Promise` in JavaScript, or `Future` in Java):

```
resultCh:=make(chan Result)
jobs<-Work{
    file:"someFile",
    pattern: compiledPattern,
    ch:resultCh,
}
```

```
// Do other things...
for result := range <-resultCh {
    ...
}
```

How is this different from directly calling the worker function? Suppose you are writing a server program, and the request includes the filename and the pattern to search for. If thousands of requests arrive concurrently, it will use thousands of open files, which may not be possible on your platform. If you use the worker pool approach, no matter how many requests arrive concurrently, there will be, at most, the predefined number of workers (and thus, that many open file). So, worker pools are a good way of limiting concurrency in a system.

In closing, have you noticed there are no mutexes in this worker implementation? And the only explicit wait is the `WaitGroup` that waits for all workers to complete?

Pipelines, fan-out, and fan-in

Many times, a computation has to go through multiple stages that transform and enrich the result. Typically, there is an initial stage that acquires a sequence of data items. This stage passes those data items one by one to successive stages, where each stage operates on the data, produces a result, and passes it on to the next stage. A good example is image processing pipelines, where the image is decoded, transformed, filtered, cropped, and encoded into another image. Many data processing applications work with large amounts of data. Therefore, a concurrent pipeline can be essential for acceptable performance.

In this chapter, we will build a simple data processing pipeline that reads records from a **comma-separated values (CSV)** text file. Each record contains a height and weight measurement for a person captured as inches and pounds. Our pipeline will convert these measurements to centimeters and kilograms, then output them as a stream of JSON objects. We will use some generic functions to abstract the staging aspects of the problem so the actual computational units do not change from one pipeline implementation to the other.

The `Record` structure is defined as follows:

```
type Record struct {
    Row     int     `json:"row"`
    Height  float64 `json:"height"`
    Weight  float64 `json:"weight"`
}
```

This pipeline has three stages:

- **Parse**: This accepts a row of data read from the file. It then parses the row number as an integer, the height and weight values as floating point numbers, and returns a `Record` structure:

```go
func newRecord(in []string) (rec Record, err error) {
    rec.Row, err = strconv.Atoi(in[0])
    if err != nil {
        return
    }
    rec.Height, err = strconv.ParseFloat(in[1], 64)
    if err != nil {
        return
    }
    rec.Weight, err = strconv.ParseFloat(in[2], 64)
    return
}

func parse(input []string) Record {
    rec, err := newRecord(input)
    if err != nil {
        panic(err)
    }
    return rec
}
```

- **Convert**: This accepts a `Record` structure as input. It then converts the height and weight to centimeters and kilograms and outputs the converted `Record` structure:

```go
func convert(input Record) Record {
    input.Height = 2.54 * input.Height
    input.Weight = 0.454 * input.Weight
    return input
}
```

- **Encode**: This accepts a `Record` structure as input. It encodes the record as a JSON object.

```go
func encode(input Record) []byte {
    data, err := json.Marshal(input)
    if err != nil {
        panic(err)
```

```
        }
        return data
    }
```

There are several ways a pipeline can be built. The most straightforward method is a synchronous pipeline. The synchronous pipeline simply passes the output of one function to the other. The input to the pipeline is read from the CSV file:

```
func synchronousPipeline(input *csv.Reader) {
    // Skip the header row
    input.Read()
    for {
        rec, err := input.Read()
        if err == io.EOF {
            return
        }
        if err != nil {
            panic(err)
        }
        // The pipeline: parse, convert, encode
        out := encode(convert(parse(rec)))
        fmt.Println(string(out))
    }
}
```

The execution of this pipeline is depicted in *Figure 5.2*. The pipeline simply processes one record to completion and then processes the subsequent one until all records are processed. If each stage takes, say, 1 μs, it produces an output every 3 μs.

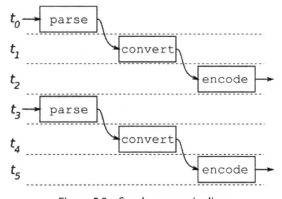

Figure 5.2 – Synchronous pipeline

Asynchronous pipeline

An asynchronous pipeline runs each stage in a separate goroutine. Each stage reads the next input from a channel, processes it, and writes it to the output channel. When the input channel closes, it closes the output channel, which causes the next stage to close its channel, and so on, until all channels are closed and the pipeline is terminated. The benefit of this type of operation is evident from *Figure 5.3*: assuming all the stages run in parallel, if each stage takes 1 µs, this pipeline produces an output every 1 µs after the initial 3 µs.

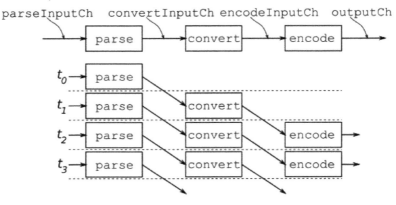

Figure 5.3 – Asynchronous pipeline

We can use some generic functions to wire the stages of this pipeline together. We wrap each stage in a function that reads from a channel in a forloop, calls a function to process the input, and writes the output to an output channel:

```
func pipelineStage[IN any, OUT any](input <-chan IN, output
chan<- OUT, process func(IN) OUT) {
    defer close(output)
    for data := range input {
        output <- process(data)
    }
}
```

Here, the IN and OUT type parameters are the input and output data types for the process function, respectively, as well as the channel types for the input and output channels.

The setup for the asynchronous pipeline is a bit more involved because we have to define a separate channel to connect each stage:

```
func asynchronousPipeline(input *csv.Reader) {
    parseInputCh := make(chan []string)
```

```go
convertInputCh := make(chan Record)
encodeInputCh := make(chan Record)
outputCh := make(chan []byte)
// We need this channel to wait for the printing of
// the final result
done := make(chan struct{})

// Start pipeline stages and connect them
go pipelineStage(parseInputCh, convertInputCh, parse)
go pipelineStage(convertInputCh, encodeInputCh, convert)
go pipelineStage(encodeInputCh, outputCh, encode)

// Start a goroutine to read pipeline output
go func() {
    for data := range outputCh {
        fmt.Println(string(data))
    }
    close(done)
}()

// Skip the header row
input.Read()
for {
    rec, err := input.Read()
    if err == io.EOF {
        close(parseInputCh)
        break
    }
    if err != nil {
        panic(err)
    }
    // Send input to pipeline
    parseInputCh <- rec
}
// Wait until the last output is printed
<-done
}
```

You may have noticed this pipeline looks like worker pools connected one after another. In fact, each stage can be implemented as a worker pool. Such a design may be useful if some of the stages take a long time to complete, so multiple of them running concurrently can increase throughput.

Fan-out/fan-in

In an ideal case where all workers are running in parallel and with two workers at each stage, the pipeline operation looks like *Figure 5.4*. If each stage can produce an output every 1 μs, this pipeline will produce two outputs every 1 μs after the initial 3 μs:

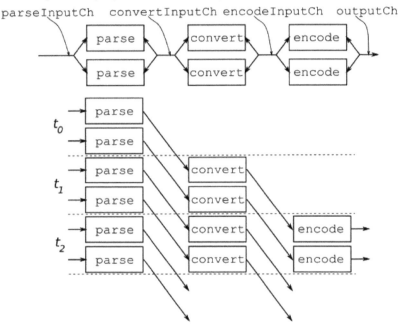

Figure 5.4 – Asynchronous pipeline with two workers at each stage

In this design, the stages of the pipeline communicate using a shared channel, so multiple goroutines read from the same `input` channel (fan-out), and they write to a shared `output` channel (fan-in).

This pipeline requires some changes to our generic function. The previous generic function relies on the closing of the `input` channel to close its own `output` channel, so the stages can shut down one after the other. With multiple instances of the worker running at each stage, each of those workers will try to close the `output` channel, causing panic. We have to close the `output` channel once all the workers terminate. So, we need a `WaitGroup`:

```
func workerPoolPipelineStage[IN any, OUT any](input <-chan IN,
    output chan<- OUT, process func(IN) OUT, numWorkers int) {
```

```
        // close output channel when all workers are done
        defer close(output)
        // Start the worker pool
        wg := sync.WaitGroup{}
        for i := 0; i < numWorkers; i++ {
                wg.Add(1)
                go func() {
                        defer wg.Done()
                        for data := range input {
                                output <- process(data)
                        }
                }()
        }
        // Wait for all workers to finish
        wg.Wait()
}
```

When the `input` channel closes, all the workers in the pipeline stage will terminate one by one. The `WaitGroup` ensures that the function does not return until all goroutines are done, and after that, it closes the `output` channel, which triggers the same sequence of events in the next stage.

The pipeline setup now uses this generic function:

```
numWorkers := 2
// Start pipeline stages and connect them
go workerPoolPipelineStage(parseInputCh, convertInputCh, parse,
numWorkers)
go workerPoolPipelineStage(convertInputCh, encodeInputCh,
convert, numWorkers)
go workerPoolPipelineStage(encodeInputCh, outputCh, encode,
numWorkers)
```

If you build this pipeline and run it, soon you will realize that the output may look like this:

```
{"row":65,"height":172.72,"weight":97.61}
{"row":64,"height":195.58,"weight":81.266}
{"row":66,"height":142.24,"weight":101.242}
{"row":68,"height":152.4,"weight":80.358}
{"row":67,"height":162.56,"weight":104.87400000000001}
```

The rows are out of order! Because there are multiple instances of data going through the pipeline, the fastest one shows up at the output first, which may not be the first one in. There are many cases where the order of records coming out of the pipeline is not important. In some cases, though, you need them in order. This pipeline construction is not a good candidate for those problems. Sure, you can add a new stage to sort them, but you'll need a potentially unbounded buffer: if there are multiple workers at each stage and if the first record takes so long to process that all the other records go through the pipeline before the first one, you have to buffer them all to sort them. That defeats the purpose of having a pipeline.

We will look at an alternate pipeline design that can deal with this problem. Our pipelines so far have used shared channels between stages that all workers send to and receive from. Another option is to use dedicated channels between the goroutines of each stage. This design becomes especially beneficial when some of the stages of the pipeline are expensive. Having multiple goroutines available to compute the expensive operations concurrently can increase the throughput of the whole pipeline.

For our example, let's suppose the conversion stage performs expensive computations, so we want to have a worker pool containing multiple workers at this stage. So, after the pipeline parses the input, it fans out to multiple conversion goroutines that read from a shared channel, but these goroutines each return their responses in their own channels. So, before we can encode the output of this stage, we have to fan-in and order the results. This is depicted in *Figure 5.5*:

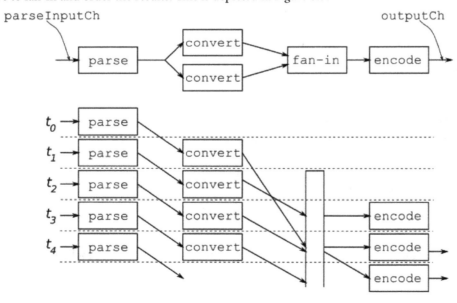

Figure 5.5 – Fan-out/fan-in without ordering

We need a new generic function that takes an input channel together with a done channel for cancellations and returns an output channel. This way we can connect the output of one goroutine to the input of another goroutine in another stage:

```
func cancelablePipelineStage[IN any, OUT any](input <-chan IN,
done <-chan struct{}, process func(IN) OUT) <-chan OUT {
    outputCh := make(chan OUT)
    go func() {
        for {
            select {
                case data, ok := <-input:
                    if !ok {
                        close(outputCh)
                        return
                    }
                    outputCh <- process(data)
                case <-done:
                    return
            }
        }
    }()
    return outputCh
}
```

Now, we can write a generic fan-in function:

```
func fanIn[T any](done <-chan struct{}, channels ...<-chan T)
<-chan T {
    outputCh := make(chan T)
    wg := sync.WaitGroup{}
    for _, ch := range channels {
        wg.Add(1)
        go func(input <-chan T) {
            defer wg.Done()
            for {
                select {
                    case data, ok := <-input:
                        if !ok {
```

```
                                        return
                                }
                                outputCh <- data
                        case <-done:
                                return
                        }
                }
        }(ch)
    }
    go func() {
        wg.Wait()
        close(outputCh)
    }()
    return outputCh
}
```

The `fanIn` function gets multiple channels, reads from each channel concurrently using separate goroutines, and writes to the common `output` channel. When all the `input` channels are closed, the receiving goroutines terminate, and the `output` channel is closed. As you can see, the output is not necessarily in the same order as the input. The goroutines may shuffle the input based on the order in which they run.

> **A side note is necessary here...**
>
> If the number of input channels is fixed, a `select` statement is an easy way to fan-in. But here, the number of input channels can be dynamic and very large. A `select` statement with a variable number of channels would be perfect for these cases. The Go language syntax does not support it, but the standard library does. The `reflect.Select` function allows you to select using a slice of channels.

The following snippet wires the stages of the pipeline with 2 workers for the conversion stage:

```
// single input channel to the parse stage
parseInputCh := make(chan []string)
convertInputCh := cancelablePipelineStage(parseInputCh, done,
parse)
numWorkers := 2
fanInChannels := make([]<-chan Record, 0)
for i := 0; i < numWorkers; i++ {
    // Fan-out
```

```
.convertOutputCh :=
    cancelablePipelineStage(convertInputCh,
    done, convert)
  fanInChannels = append(fanInChannels, convertOutputCh)
}
convertOutputCh := fanIn(done, fanInChannels...)
outputCh := cancelablePipelineStage(convertOutputCh, done,
encode)
```

Fan-in with ordering

How can we write a fan-in function that can also order the records? The key idea is to store the out-of-order records until the expected record comes. Let's say we have two input channels, listened to by two goroutines, and the first goroutine receives an out-of-order record. We know that the second goroutine will receive the expected record next time because the number of records in the pipeline cannot exceed the number of concurrent workers, and we already received the second record, so the first record is still in the previous stages. While waiting for that record to arrive, we have to prevent the first goroutine from returning another record. But how can we pause a running goroutine? The answer is: by making it wait on a channel.

Let's try to put together an algorithm in pseudocode. I find it helpful to write pseudocode blocks representing goroutines and draw arrows between them to denote message exchanges. For each input channel, we will use a goroutine that receives the data element from the pipeline, sends it to a fan-in channel that the ordering goroutine is receiving from, and waits to receive from a pause channel. In the second stage, we have the ordering goroutine that receives from the fan-in channel and determines whether the record is in the correct order. If not, it stores this order at a buffer dedicated to its input channel. At this point, the goroutine for that input channel is waiting to receive from its pause channel, so it cannot accept any more inputs. When the correct input comes, the ordering goroutine outputs all queued data and releases all waiting goroutines by sending them to their pause channels. This is illustrated in *Figure 5.6*:

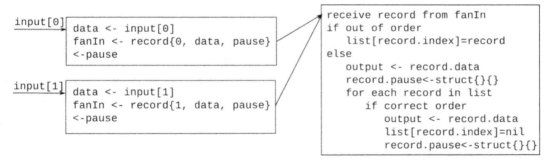

Figure 5.6 – Pseudocode for ordering fan-in

So, let's start constructing this ordered fan-in algorithm. First, we need a way to get the sequence number of records:

```
type sequenced interface {
    getSequence() int
}
func (r Record) getSequence() int { return r.Row }
```

For each channel, we need a place to store the out-of-order records and a channel to pause the goroutine:

```
type fanInRecord[T sequenced] struct {
    index int // index of the input channel
    data  T
    pause chan struct{}
}
```

We create a goroutine for each input channel. Each goroutine reads from its assigned channel, creates an instance of fanInRecord, and sends it via the fanInCh channel. This may be the expected record, or it may be out-of-order, but that is up to the receiving end of the fanInCh channel. This goroutine now has to pause until that determination is made. So, it receives from the associated pause channel. Another goroutine releases the goroutine by sending a signal to the pause channel, after which the goroutine starts listening to the input channel again. Of course, if the input channel is closed, the corresponding goroutine returns, and when all the goroutines return, the fanInCh channel closes:

```
func orderedFanIn[T indexable](done <-chan struct{}, channels
...<-chan T) <-chan T {
  fanInCh := make(chan fanInRecord[T])
  wg := sync.WaitGroup{}
  for i := range channels {
    pauseCh := make(chan struct{})
    wg.Add(1)
    go func(index int, pause chan struct{}) {
      defer wg.Done()
      for {
        var ok bool
        var data T
        // Receive input from the channel
        select {
          case data, ok = <-channels[index]:
            if !ok {
```

```
                    return
                }
                // Send the input
                fanInCh <- fanInRecord[T]{
                    index: index,
                    data:  data,
                    pause: pause,
                }
            case <-done:
                return
            }
            // pause the goroutine
            select {
                case <-pause:
                case <-done:
                    return
            }
        }
    }(i, pauseCh)
}
go func() {
        wg.Wait()
        close(queue)
    }()
```

The second part of the function contains the ordering logic. When an out-of-order record is received from a channel, it is stored in the buffer dedicated to that channel, so we only need a buffer of len(channels) capacity. When the expected record is received, the algorithm scans the stored records and outputs them in the correct order:

```
outputCh := make(chan T)
go func() {
    defer close(outputCh)
    // The next record expected
    expected := 1
    queuedData := make([]*fanInRecord[T], len(channels))
    for in := range fanInCh {
```

```go
          // If this input is what is expected, send
          //it to the output
          if in.data.getSequence() == expected {
            select {
            case outputCh <- in.data:
              in.pause <- struct{}{}
              expected++
              allDone := false
              // Send all queued data
              for !allDone {
                allDone = true
                for i, d := range queuedData {
                  if d != nil && d.data.getSequence() ==
                  expected {
                    select {
                    case outputCh <- d.data:
                      queuedData[i] = nil
                      d.pause <- struct{}{}
                      expected++
                      allDone = false
                    case <-done:
                      return
                    }
                  }
                }
              }
            case <-done:
              return
            }
          } else {
            // This is out-of-order, queue it
            in := in
            queuedData[in.index] = &in
          }
        }
      }()
```

```
    return outputCh
}
```

The idea of this goroutine is to listen to the queue channel, and if the received record is out-of-order, queue it. The sending goroutine will be blocked until this goroutine releases it. If the correct record comes, it is directly sent to the output channel, the goroutine that sent it is unblocked, and all queued records are scanned to see if the next expected record is already queued. If so, that record is sent, the corresponding goroutine is unblocked by a send to the `pause` channel, and the record is dequeued.

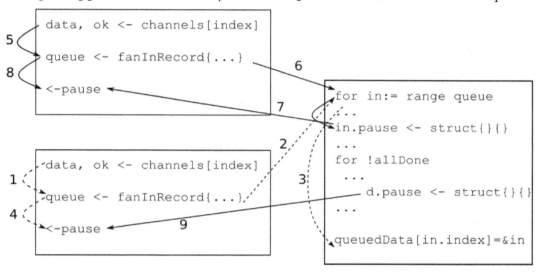

Figure 5.7 – Ordered fan-in happened-before relationships

Let's look at how these goroutines interact, which is depicted in *Figure 5.7*. One of the goroutines reads an out-of-order input (arrow 1) and sends it to the fan-in goroutine through the `fanInCh` channel (arrow 2). The fan-in goroutine, realizing this is an out-of-order record, queues it (arrow 3). While these are happening, the goroutine starts waiting to receive from its `pause` channel (arrow 4). Concurrently, another goroutine receives another input (arrow 5) and sends it to the fan-in goroutine via the `fanInCh` channel (arrow 6). The fan-in goroutine, realizing this is the expected packet, releases the goroutine (arrow 7), which is already waiting or will soon wait to receive from its `pause` channel. The fan-in goroutine also looks at the stored requests and sees that there is one record waiting, which now becomes the expected packet. So, it releases that goroutine as well (arrow 9).

As you can see, pipelines can get complicated based on the exact needs. There is no single way to solve these problems. These examples show several ways to abstract away the underlying complexities of building and running high-performance data pipelines so that the actual processing stages can concentrate on only data processing. As I tried to illustrate in this chapter, different types of pipelines can be constructed and connected using concurrent components and generic functions without a need to change the processing logic. Start simple, profile your programs, find the bottlenecks, and then you can decide if and when to fan out, how to fan in, how to size worker pools, and what type of pipeline works best for your use case.

Finally, note that all pipeline implementations here used channels, goroutines, and waitgroups. There are no critical sections, no mutexes or condition variables, and yet, no data races. Each goroutine makes progress as soon as new data becomes available.

Summary

In this chapter, we studied worker pools and pipelines – two patterns that show up in different shapes and forms in almost every non-trivial project. There are many ways these patterns can be implemented with different runtime behaviors. You should build your systems so that they do not rely on the exact structure of the pipeline or the worker pools. I tried to show some ways to abstract away concurrency concerns from computation logic. These ideas may make your job easier when you need to iterate among different designs.

Next, we will talk about error handling and how error handling can be added to these patterns.

Questions

1. Can you change the worker implementation so that the submitted work can be canceled by the caller?

2. Many languages offer frameworks with dynamically sized worker pools. Can you think of a way to implement that in Go? Would that worker pool be more performant than a fixed-sized worker pool that uses the same number of goroutines as the maximum for the dynamically sized one?

3. Try writing a generic fan-in/fan-out function (without ordering) that takes n input channels and m output channels.

6

Error Handling

This chapter is about dealing with errors and panics in a concurrent program. First, we will look at how error handling can be incorporated into concurrent programs, including how to *pass* errors between goroutines so that you can handle or report them. Then we will talk about panics.

There are no hard and fast rules about dealing with errors and panics, but I hope some of the guidelines described in this chapter will help you write more robust code. The first guideline is this: **never ignore errors**. The second guideline tells you when to return an error and when to panic: **the audience for errors is the users of the program; the audience for panics is the developers of the program**.

This chapter contains the following sections:

- Error handling
- Panics

At the end of this chapter, you will have seen several approaches to error handling in concurrent programs.

Error handling

Error handling in Go has been a polarizing issue. Frustrated with the repetitive error-handling boilerplate, many Go users in the community (including me) suggested *improved* error-handling mechanisms. Most of these proposals were actually error-passing improvements because, to be honest, errors are rarely handled. Rather, they are *passed* to the caller, sometimes wrapped with some contextual information.

A good number of error-handling proposals suggested different variations on throw-catch, while many others were simply what is called *syntactic sugar* for if err!=nil return err. Many of these proposals missed the point that the existing error reporting and handling conventions work nicely in a concurrent environment, such as the ability to pass errors via a channel: you can handle errors generated by one goroutine in another goroutine.

An important point I like to emphasize when working with Go programs is that most of the time, they can be analyzed by solely relying on the information you see on screen. All possible code paths are explicit in the code. In that sense, Go is an extremely reader-friendly language. The Go error handling paradigm is partly responsible for this. Many functions return some values combined with an error. So, how the program deals with errors is usually explicit in the code.

Errors are generated by the program when an unacceptable situation is detected, such as a failing network connection, or an invalid user input. These errors are usually passed to the caller, sometimes wrapped in another error to add information describing the context. The added information is important because many of these errors are converted into messages for the users of the program. For example, a message complaining about a JSON parsing error is useless in a program that works with many JSON files unless it also tells which file has the error.

But goroutines do not return errors. We have to find other ways to deal with errors when goroutines fail. When multiple goroutines are used for different parts of a computation, and one of the goroutines fails, the remaining goroutines should also be canceled, or their results thrown away. Sometimes, multiple goroutines fail, and you have to deal with multiple error values. There are few guidelines and many third-party packages that will make error handling easy for you. The first guideline is: never ignore errors.

Let's look at some common patterns. If you submit work to a goroutine and expect to receive a result later, make sure that result includes error information in it. The pattern illustrated here is useful if you have multiple tasks that can be performed concurrently. You start each task in its own goroutine and then collect the results or errors as needed. This is also a good way of dealing with errors when you have a worker pool:

```go
// Result type keeps the expected result, and the
// error information.
type Result1 struct {
    Result ResultType1
    Error err
}
type Result2 struct {
    Result ResultType2
    Error err
}
...
result1Ch:=make(chan Result1)
go func() {
    result, err := handleRequest1()
    result1Ch <- Result1{ Result: result, Error: err }
}()
result2Ch:=make(chan Result2)
go func() {
```

```
        result, err := handleRequest2()
        result2Ch <- Result2{ Result: result, Error: err }
}()
// Do other work
...
// Collect the results from the goroutines
result1:=<-result1Ch
result2:=<-result2Ch
if result1.Error!=nil {
// handle error
        ...
}
if result2.Error!=nil {
// handle error
        ...
}
```

You have to be mindful of the active goroutines when you detect an error. For example, the preceding program reads from all the result channels before checking for errors. This ensures that all the goroutines that were started terminate, and then error handling is performed. The following implementation would leak the second goroutine:

```
result1:=<-result1Ch
if result1.Error!=nil {
// result2Ch is never read. Goroutine leaks!
        return result1.Error
}
result2:=<-result2Ch
...
```

In many cases, it may be just fine to let all goroutines complete and then return the error, or return a composite error if multiple goroutines failed. But sometimes, you may want to cancel running goroutines if another goroutine fails. In *Chapter 8*, we will talk about using context.Context to cancel such computations. For now, we can use a canceled channel to notify the goroutines that they should stop processing. If you remember, this is a common pattern where the closing of a channel is used to broadcast a signal to all the goroutines. So, when a goroutine detects an error, it will close the canceled channel. All goroutines will periodically check whether the canceled channel is closed and return an error if so. But there is a problem with this approach: if more than one goroutines fail, they will all try to close the channel, and closing an already closed channel will panic. So instead of closing the `canceled` channel, we will have a separate goroutine that listens to a cancel channel, and closes the `canceled` channel only once:

```go
// Separate result channels for goroutines
resultCh1 := make(chan Result1)
resultCh2 := make(chan Result2)
// canceled channel is closed once when a goroutine
// sends to cancelCh
canceled := make(chan struct{})
// cancelCh can receive many cancellation requests,
// but closes canceled channel once
cancelCh := make(chan struct{})
// Make sure cancelCh is closed, otherwise the
// goroutine that reads from it leaks
defer close(cancelCh)
go func() {
    // close canceled channel once when received
    // from cancelCh
    once := sync.Once{}
    for range cancelCh {
        once.Do(func() {
            close(canceled)
        })
    }
}()
// Goroutine 1 computes Result1
go func() {
    result, err := computeResult1()
    if err != nil {
        // cancel other goroutines
        cancelCh <- struct{}{}
        // Send error back. Do not close channel
        resultCh1 <- Result1{Error: err}
        return
    }
    // If other goroutines failed, stop computation
    select {
```

```
            case <-canceled:
                    // close resultCh1, so the listener does
                    // not block
                    close(resultCh1)
                    return
        default:
        }
        // Do more computations
}()
// Goroutine 2 computes Result2
go func() {
    ...
}()
// Receive results. The channel will be closed if
// the goroutine was canceled (ok will be false)
result1, ok1 := <-resultCh1
result2, ok2 := <-resultCh2
```

Here, if goroutine 1 fails, resultCh1 will return the error, goroutine 2 will be canceled, and resultCh2 will be closed. If goroutine 2 fails, resultCh2 will return the error, goroutine 1 will be canceled, and resultCh1 will be closed. If they both fail concurrently, both errors will be returned.

A variation of this is using an error channel instead of a cancel channel. A separate goroutine listens to the error channel and captures the errors from the goroutines:

```
// errCh will communicate errors
errCh := make(chan error)
// Any error will close canceled channel
canceled := make(chan struct{})
// Ensure error listener terminates
defer close(errCh)
// collect all errors.
errs := make([]error, 0)
go func() {
    once := sync.Once{}
    for err := range errCh {
            errs = append(errs, err)
            // cancel all goroutines when error received
```

```
        once.Do(func() { close(canceled) })
    }
}()
resultCh1 := make(chan Result1)
go func() {
    defer close(resultCh1)
    result, err := computeResult()
    if err != nil {
        errCh <- err
        // Make sure listener does not block
        return
    }
    // If canceled, stop
    select {
    case <-canceled:
        return
    default:
    }
    resultCh1 <- result
}()
result, ok := <-resultCh1
```

Yet another error handling approach that I often see in the field is using dedicated error variables in the enclosing scope for each goroutine. This approach needs a WaitGroup, and there is no way to cancel work when one of the goroutines fails. Nevertheless, it can be useful if none of the goroutines perform cancelable operations. If you end up implementing this pattern, make sure errors are read after the wait group's Wait() call because, according to the Go Memory Model, the setting of the error variables happens before the return of that Wait() call, but they are concurrent until then:

```
wg := sync.WaitGroup{}
wg.Add(2)
var err1 error
go func() {
    defer wg.Done()
    if err := doSomething1(); err != nil {
        err1 = err
        return
    }
```

```
}()
var err2 error
go func() {
      defer wg.Done()
      if err := doSomething2(); err != nil {
            err2 = err
            return
      }
}()
wg.Wait()
// Collect results and deal with errors here
if err1 != nil {
      // handle err1
}
if err2 != nil {
      // handle err2
}
```

Pipelines

There are several options for error handling when working with an asynchronous pipeline. A pipeline is usually constructed to process many inputs. Because of this, it is usually not desirable to stop the pipeline just because processing failed for one of the inputs. Instead, you log or record the error and continue processing. The important thing is to capture enough context with the error so that after everything is said and done, you can go back and figure out what went wrong for what input. The options to deal with errors in a pipeline include, but are not limited to, the following:

- Each stage handles errors itself by using an error recorder function. The error recorder must be able to deal with concurrent calls if multiple stages attempt to record errors concurrently.

- Use a separate error channel with an error listener goroutine. When an error is detected in the pipeline, the relevant context is captured (the input filename, identifier, or the complete input, what went wrong, which stage failed, etc.) and sent to a channel. The error listener goroutine stores the error information in a database or logs it.

- Pass the error to the next stage. Each stage checks whether the input contains errors and passes it along until the end of the pipeline, where an error output is produced.

Servers

When I talk about servers, I mainly talk about their request-oriented nature and not about their communication characteristics. The requests may come from a network via HTTP or gRPC or they can come from the command line. Usually, each request is handled in a separate goroutine. Thus, it is up to the request-handling stack to propagate meaningful errors that can be used to build a response to the user. If that user is another program (that is, if we're talking about a web service, for instance), it makes sense to include an error code and some diagnostic message. Structured errors are your best friend:

```go
// Embed this error to all other structured errors that can be
returned from the API
type Error struct {
    Code int
    HTTPStatus int
    DiagMsg string
}
// HTTPError extracts HTTP information from an error
type HTTPError interface {
    GetHTTPStatus() int
    GetHTTPMessage() string
}

func (e Error) GetHTTPStatus() int {
  return e.HTTPStatus
}

func (e Error) GetHTTPMessage() string {
    return fmt.Sprintf("%d: %s",e.Code,e.DiagMsg)
}

// Treat HTTPErrors and other unrecognized errors
//separately
func WriteError(w http.ResponseWriter, err error) {
    if e, ok:=err.(HTTPError); ok {
        http.Error(w,e.HTTPStatus(),e.HTTPMessage())
    } else {
        http.Error(w,http.InternalServerError,err.Error())
    }
}
```

Error implementations such as the preceding one will help you return meaningful errors to your users, so they can deal with common problems without wasting hours in frustration.

Panics

Panics are different from errors. A panic is either a programming error or a condition that cannot be reasonably remedied (such as running out of memory.) Because of this, a panic should be used to convey as much diagnostic information to the developer as possible.

Some errors can become panics depending on the context. For instance, a program may accept a template from the user and generate an error if the template parsing fails. However, if the parsing of a hardcoded template fails, then the program should panic. The first case is a user error, and the second case is a bug.

As a developer of concurrent programs, there are only three things you can do with an error: you either handle it (log it, choose another program flow, or ignore it by doing nothing), you pass it to the caller (sometimes with some additional contextual information), or you panic. When a panic happens in a concurrent program, the runtime ensures that all the nested function calls return, one by one, all the way to the function that started that goroutine. While this is happening, all deferred blocks of the functions also run. This is a chance to recover from a panic, or to clean up any resources that will not be garbage-collected. If the panic is not handled by one of the functions in the call chain, the program will crash. So, as the developer, you have some cleaning up to do.

In a server program, usually, a separate goroutine handles each request. Most server frameworks (including the standard library net/http package) handle such panics without crashing by printing a stack and failing the request. If you are writing a server without using such a library or if you want to report more information when you catch a panic, you should handle them yourself:

```
func PanicHandler(next func(http.ResponseWriter,*http.Request))
func(http.ResponseWriter,*http.Request) {
   return func(wr http.ResponseWriter, req *http.Request) {
     defer func() {
          if err:=recover(); err!=nil {
             // print panic info
          }
     }()
     next(wr,req)
   }
}

func main() {
```

```
        http.Handle("/path",PanicHandler(pathHandler))
    }
```

You can only recover a panic in the goroutine if it is initiated. That means if you start a goroutine that can initiate a panic and you do not want that panic to terminate the program, you have to recover:

```
    go func(errCh chan<- error, resultCh chan<- result) {
        defer func() {
            if err:=recover(); err!=nil {
                // panic recovered, return error instead
                errCh <- err
            close(resultCh)
            }
        }()
        // Do the work
    }()
```

When working with concurrent processing pipelines (such as the ones we worked on in *Chapter 5*), it makes sense to deal with panics defensively. A panic usually points to a bug in the program but terminating a long-running pipeline after hours of processing is not the best solution. You usually want to have a log of all the panics and errors once the processing is complete. So, you have to make sure that the panic recovery is performed at the correct place. For instance, in the following code snippet, the panic recovery is around the actual pipeline stage processing function, so a panic is recorded, but the for loop continues processing:

```
    func pipelineStage[IN any, OUT WithError](input <-chan IN,
    output chan<- OUT, errCh chan<-error, process func(IN) OUT) {
        defer close(output)
        for data := range input {
            // Process the next input
            result, err := func() (res OUT,err error) {
                defer func() {
                    // Convert panics to errors
                    if err = recover(); err != nil {
                        return
                    }
                }()
                return process(data),nil
            }()
```

```
        if err!=nil {
                // Report error and continue
                errCh<-err
                continue
        }
        output<-result
    }
}
```

If you are familiar with exception-handling mechanisms in C++ or Java, you might wonder whether panics can be used in place of throwing an exception. Some guidelines strongly discourage that, but you can find other resources advocating the exact same thing. I will leave that judgment to you, but there are examples of it in the standard library JSON package as well, and one might argue that if you have a large package with very few exported functions, it may make sense to use panics as an error-handling mechanism because it becomes an implementation detail. JSON unmarshaling is one example, and a deeply nested parser is another case where this might help. If you decide to do it, here's the way: use package-level error types to distinguish between real panics and errors. The following snippet is a modified version of the standard library JSON unmarshaling implementation:

```
// All internal functions panic with this type of
// error instead of returning an error
type packageError struct{ error }


// Exported function is the top-level function that
// calls the unexported implementation functions and
// recovers panics
func ExportedFunction() (err error) {
    defer func() {
            if r := recover(); r != nil {
            // If the panic is an error thrown from the
            // package recover and return error
                if e, ok := r.(packageError); ok {
                        err = e.error
                } else {
                        // This is a real panic
                        panic(r)
                }
            }
    }
```

```
        } ()
        unexportedFunction()
        return nil
    }


    // unexportedFunction is the top-level of the
    // implementation
    func unexportedFunction() {
        if err:=doThings(); err!=nil {
            panic(packageError{err})
        }
        ...
    }
```

Here, `unexportedFunction` performs the actual work, and `ExportedFunction` acts as the external interface of `unexportedFunction` by translating some of the panics into errors.

Summary

Your programs must generate useful error messages that tell the users what went wrong and how they can fix it. Go gives the developer full control over how errors are generated and how they are passed around. In this chapter, we saw some methods to deal with errors generated concurrently.

Next, we will look at timers and tickers for scheduling events in the future.

7
Timers and Tickers

Many long-lived applications impose limits on how long an operation can last. They also perform tasks such as health checks periodically to ensure all components are working as expected. Many platforms provide high-precision timer operations, and the Go standard library provides portable abstractions of these services in the time package. We will look at timers and tickers in this chapter. Timers are tools for doing things later, and tickers are tools for doing things periodically.

The key sections we will review in this chapter are the following:

- Timer – running something later
- Tickers – running something periodically

At the end of this chapter, you will have seen how to work with timers and tickers and how you can monitor other goroutines using heartbeats.

Technical Requirements

The source code for this particular chapter is available on GitHub at https://github.com/PacktPublishing/Effective-Concurrency-in-Go/tree/main/chapter7.

Timers – running something later

If you want to do something later, use time.Timer. A Timer is a nice way of doing the following:

```
// This is only for illustration. Don't do this!
type TimerMockup struct {
    C chan<- time.Time
}

func NewTimerMockup(dur time.Duration) *TimerMockup {
    t := &TimerMockup{
```

```
                C: make(chan time.Time,1),
        }
        go func() {
                // Sleep, and then send to the channel
                time.Sleep(dur)
                t.C <- time.Now()
                }()
        return t
}
```

So, a timer is like a goroutine that will send a message to a channel after sleeping for the requested amount of time. The actual implementation of Timer uses platform-specific timers, so it is more accurate and not as simple as starting a goroutine and waiting. One thing to keep in mind is that when you receive the event from a timer channel, it means the timer duration elapsed *when the message was sent*, which is not the same as when the message was received.

You might have noticed that the timer uses a channel with a capacity of 1. This prevents goroutine leaks if the timer channel is never listened to by another goroutine. A buffered channel means the event will be generated when the duration elapses, but if no goroutines are listening to the channel, the event will wait in the channel until it is read or the timer is garbage-collected.

A common use of a timer is to limit the running time of tasks:

```
func main() {
        // timer will be used to cancel work after 100 msec
        timer := time.NewTimer(100 * time.Millisecond)
        // Close the timeout channel after 100 msec
        timeout := make(chan struct{})
        go func() {
                <-timer.C
                close(timeout)
                fmt.Println("Timeout")
        }()
        // Do some work until it times out
        x := 0
        done := false
        for !done {
                // Check if timed out
                select {
```

```
        case <-timeout:
                done = true
        default:
        }
        time.Sleep(time.Millisecond)
        x++
    }
    fmt.Println(x)
}
```

The timer setup can be greatly simplified by using the `time.AfterFunc` function. The following function call can replace the timer setup and goroutine in the preceding code snippet. The `time.AfterFunc` function will simply call the given function after the given duration:

```
time.AfterFunc(100*time.Millisecond, func() {
    close(timeout)
    fmt.Println("Timeout") })
```

A similar approach would be to use `time.After`:

```
ch := time.After(100*time.Millisecond)
```

Then, the `ch` channel will receive a time value after `100` milliseconds.

Stopping a timer is easy. In the preceding program, if the long-running task finishes before it times out, we want to stop the timer; otherwise, it will print out an erroneous `Timeout` message. A call to `Stop()` may manage to stop the timer if it hasn't expired yet, or the timer may expire after you call `Stop()`. These two cases are illustrated in *Figure 7.1*.

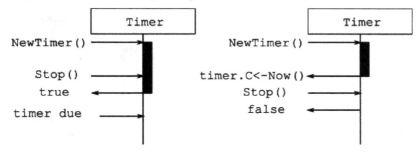

Figure 7.1 – Stopping a timer before and after it fires

If `Stop()` returns `true`, then you managed to stop the timer. However, if `Stop()` returns `false`, the timer expired, and thus, it has already stopped. This doesn't mean that the message from the timer channel has been consumed yet and may be consumed after Stop() returns. Do not forget that the timer channel has a capacity of 1, so the timer will send to that channel even if nobody is receiving from it.

The Timer type allows you to reset the timer. The behavior is different between a timer created by NewTimer, and a timer created by AfterFunc, as follows:

- If the timer is created by AfterFunc, resetting the timer will either reset the time the function will run for the first time (in which case, Reset will return true) or it will set the time the function will run one more time (in which case, Reset will return false).

- If the timer is created by NewTimer, resetting can only be done on a stopped and drained timer. Also, draining and resetting a timer cannot be concurrent with the goroutine that receives from the timer. The correct way of doing this is shown in the following code block. The important point to note here is that while timer draining and resets happen, it is not possible to receive from the timer channel using the timeout case of the select statement. In other words, while resetting a timer, no other goroutine should be listening from that timer's channel:

```
select {
    case <-timer.C:
    // Timeout
    case d:=<-resetTimer:
        if !timer.Stop() {
            <-timer.C
        }
    timer.Reset(d)
}
```

Different and interesting use cases for timers and especially AfterFunc come up often. For timeouts, context.Context is a more idiomatic tool. We will look at that in the next chapter.

Tickers – running something periodically

It may be a reasonable idea to run a function periodically using repeated calls to AfterFunc:

```
var periodicTask func()
periodicTask = func() {
    DoSomething()
    time.AfterFunc(time.Second, periodicTask)
}
time.AfterFunc(time.Second, periodicTask)
```

With this approach, each run of the function will schedule the next one, but variations in the running duration of the function will accumulate over time. This may be perfectly acceptable for your use case, but there is a better and easier way to do this: use time.Ticker.

time.Ticker has an API very similar to that of time.Timer: You can create a ticker using time.NewTicker, and then listen to a channel that will periodically deliver a tick until it is explicitly stopped. The period of the tick will not change based on the running time of the listener. The following program prints the number of milliseconds elapsed since the beginning of the program for 10 seconds:

```go
func main() {
    start := time.Now()
    ticker := time.NewTicker(100 * time.Millisecond)
    defer ticker.Stop()
    done := time.After(10 * time.Second)
    for {
        select {
        case <-ticker.C:
            fmt.Printf("Tick: %d\n",
                time.Since(start).Milliseconds())
        case <-done:
            return
        }
    }
}
```

What happens if you cannot finish the task before the next tick arrives? Should you worry about receiving a bunch of ticks if you miss several of them? Fortunately, time.Ticker deals with these situations reasonably. Let's assume we have a task that we trigger using a ticker that may or may not finish before the next tick arrives. This could be a network call to a third-party service that takes longer than expected or a database call under heavy load. Whatever the reason, when the next tick arrives, the task is not ready to receive it because the task is not yet been finished.

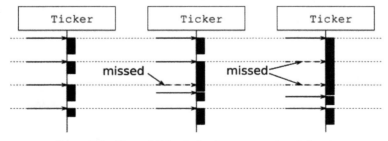

Figure 7.2 – Normal ticker behavior versus missed signals

The behavior of `Ticker` in this situation is illustrated in *Figure 7.2*. In the leftmost diagram, the task finishes consistently before the next tick arrives, so execution is periodic with uniform intervals. The middle diagram shows a situation where the first execution of the task is completed before the next tick arrives, but the second execution takes longer, and the application misses the tick. In this case, the next tick simply arrives as soon as the application listens to the channel. The third execution starts later than usual, but the fourth execution recovers the regular rhythm. The rightmost diagram shows a situation where the first execution of the task takes so long that multiple ticks are missed. When this happens, the next tick arrives as soon as the task listens to the channel, and subsequent ticks arrive at the regular rhythm. In short, at most, one message is waiting in the ticker channel. If you miss multiple ticks, you only receive one tick for those missed ticks.

An important point to remember is that you must stop tickers when you are done with them using the `Stop()` method. Unlike a `Timer` that will fire once and then be garbage-collected, a `Ticker` has a goroutine that continuously sends ticks via a channel, and if you forget to stop the ticker, that goroutine will leak. It will never be garbage-collected. Instead, use `defer ticker.Stop()`.

Heartbeats

A timeout is useful to limit the execution time of a function. When that function is expected to take a long time to return, or it does not return at all, timeouts don't work. You need a way to monitor that function to ensure that it is making progress and that it is still alive. There are several ways this can be done.

One way of doing so is by writing a long-running function to report its progress to a monitor. These reports do not have to arrive uniformly. If the monitor realizes that the long-running function has not been reported for a while, it can attempt to stop the process, alert the administrator, or print an error message. Such a monitoring function is given in the following code block. This function expects to receive information from the `heartbeat` channel from the long-running function. If a `heartbeat` signal does not arrive between two consecutive timer ticks, the process is assumed to be dead, and the `done` channel is closed in an attempt to cancel the process:

```go
func monitor(heartbeat, done chan struct{}, tick <-chan time.
Time) {
    // Keep the time last heartbeat is received
    var lastHeartbeat time.Time
    var numTicks int
    for {
        select {
            case <-tick:
                numTicks++
                if numTicks >= 2 {
                    fmt.Printf("No progress since
```

```
                                %s, terminating\n",
                                lastHeartbeat)
                                close(done)
                                return
                    }
            case <-heartbeat:
                    lastHeartbeat = time.Now()
                    numTicks = 0
        }
    }
}
```

The long-running function has the following general structure:

```
func longRunningProcess(heartbeat, done chan struct{}) {
    for {
            // Do something that can take a long time
            DoSomething()
            select {
                case <-done:
                        return
                case heartbeat <- struct{}{}:
                        // This select statement can have a
                        // default case for
                        // non-blocking operation
            }
    }
}
```

The ticker determines the maximum allowed duration for the long-running function to remain quiet:

```
func main() {
    heartbeat := make(chan struct{})
    done := make(chan struct{})
    // Expect a heartbeat at least every second
    ticker := time.NewTicker(time.Second)
    defer ticker.Stop()
    go longRunningProcess(heartbeat, done)
```

```
    go monitor(heartbeat, done, ticker.C)
    <-done
}
```

This heartbeat implementation simply sends a `struct{}{}` value. It can also send an increasing sequence of values to show progress or other types of metadata so that a progress indication can be logged or displayed to the end user.

There is no guarantee that a hung goroutine will have the chance to read from the done channel and gracefully return. It may just sit there waiting for an event that will never happen, with no indication of progress. This is especially relevant for third-party libraries or APIs over which you have no control. There isn't much you can do in that case. You can close the done channel and hope that the goroutine will eventually terminate. You should, however, log such occurrences so they can be dealt with outside the program. I have seen instances where such situations are handled by placing the process that cannot be terminated in a separate binary. The second binary performs the long-running tasks, and after a while, it dies because of the unfixable resource leak. It is brought up again either by the orchestration software or by the program itself.

Summary

Timers and tickers allow you to do things in the future and do things periodically. We only looked at a few use cases here. They are versatile tools that show up quite often in unexpected places. The Go runtime provides extremely efficient implementations of these tools. You need to be careful, though because they invariably complicate the flow. Make sure to close your tickers.

In the remaining chapters, we will start putting things together and look at some real-life use cases for concurrency patterns.

8

Handling Requests Concurrently

Server programming is a rather large topic. This chapter will mainly focus on some of the concurrency-related aspects of server programming and, in an abstract sense, request handling in general. At the end of the day, almost all programs are written to handle particular requests. For a server application, defining and propagating a request context is very important, so we start the chapter by talking about the context package. Next, we will look at some simple servers to explore how requests can be handled concurrently, and discuss some methods to deal with a few basic problems of server development. The last part of the chapter is on streaming data, where data elements are generated piecemeal, which poses unique challenges to demonstrate some interesting concurrency patterns.

This chapter includes the following sections:

- The context, cancelations, and timeouts
- Backend services
- Streaming data

By the end of this chapter, you should have a good understanding of request contexts, how you can cancel or timeout requests, building blocks of server programming, ways to limit concurrency, and how to deal with data that is generated piecemeal.

Technical Requirements

The source code for this particular chapter is available on GitHub at `https://github.com/PacktPublishing/Effective-Concurrency-in-Go/tree/main/chapter8`.

The context, cancelations, and timeouts

In *Chapter 2*, we showed that closing a channel shared between multiple goroutines is a good way to signal cancelation. Cancelations may happen in different ways: a failure in a part of the computation may invalidate the entire result, the computation may last so long that it times out, or the requester notifies the server application that it is no longer interested in the result by closing the network connection. So, it makes sense to pass a channel to the functions that are called to handle a request. But you have to be careful: you can close a channel only once. Closing a closed channel will panic. Here, the term "request" should be taken in an abstract sense: it can be an API request submitted to a server, or it can simply be a function call to handle a particular piece of a larger computation.

It also makes sense to let the functions in the call chain know about certain data related to the request. For example, in a concurrent system with many goroutines, it is important to correlate log messages with requests, so the fulfillment of a request can be traced across different goroutines. A unique request identifier is a common way of achieving this. To accommodate this, all the functions that are called by those services should know about this request identifier.

A `context.Context` is an object that deals with these two common issues. It includes the `Done` channel we talked about for cancelations, and it behaves like a generic key-value store. A `Context` is specifically designed to be used as a request-scoped object. It is a good place to store request identifiers, caller identity and privilege information, a request-specific logger, and so on.

A common mistake for those who are familiar with other languages is to treat context as a thread-local storage. Context is **not** a replacement for thread-local variables; they are meant to be shared among goroutines handling a request. Context instances cannot cross process boundaries; for example, when you call an HTTP API, the server request handler starts with a brand-new context that has no relation to the client context used to make that call. It *should* be passed as the first argument for the functions that need it. Following these conventions will make it easier to understand the code for the readers and allow static code analyzers to produce more sensible reports.

Creation and preparation of a context is usually the first order of business when handling a request. Create a new context using the following:

```
ctx := context.Background()
```

This will create an empty context with no cancelation or timeout. The `Done` channel of the context will be `nil`, thus it is not cancelable.

You work with a context by adding features to it (ever heard of the "decorator pattern"?). When you add a cancelation or timeout feature to a context, you get a new context that wraps the original context you passed in:

```
ctx1, cancel1 := context.WithCancel(ctx0)
```

Here, `ctx1` is a new context that refers to the original context, `ctx0`, but with an added cancelation feature. You pass `ctx1` to the functions and goroutines that support canceling by checking the `ctx1.Done()` channel. When you call the `cancel1` function, it will close the `ctx1.Done()` channel, so all goroutines and functions checking for the `ctx1.Done()` channel will receive that cancelation request. You can call a cancelation function many times; the underlying channel will be closed only the first time you call it. If the original context, `ctx0`, already had a cancelation feature added to it, it will not be affected by the cancelation of `ctx1`. However, `ctx1` will be canceled if `ctx0` is canceled. If other cancelable contexts are created based on `ctx1`, those contexts will be canceled whenever `ctx1` is canceled, but `ctx1` will not know about the cancelations of those nested contexts. The proper use of the cancelation feature is as follows:

```
 1: func someFunc(ctx context.Context) error {
 2:     ctx1, cancel1 := context.WithCancel(ctx)
 3:     defer cancel1()
 4:     wg:=sync.WaitGroup{}
 5:     wg.Add(1)
 6:     go func() {
 7:         defer wg.Done()
 8:         process2(ctx1)
 9:     }()
10:     if err:=process1(ctx1); err!=nil {
11:         cancel1()
12:         return err
13:     }
14:     wg.Wait()
15:     return nil
16: }
```

This function calls two separate functions, `process1` and `process2`, to perform some computation. The `process2` function is called in a separate goroutine. If `process1` fails, we want to cancel `process2`. To do this, we create a cancelable context (line 2) and make sure that this new context is canceled when this function returns (line 3). This is necessary to prevent goroutine leaks because, as you may guess, additional goroutines are necessary to implement such cascading cancelation. The call to cancel a function ensures those goroutines are terminated.

This situation is illustrated in *Figure 8.1*. `ctx0` is the initial context, with a `nil` Done channel. `ctx1` is a cancelable context created from `ctx0`, thus `cancel1` is a closure that closes the done channel for `ctx1`. There is no need for an additional goroutine as the parent context was not cancelable. `ctx2` is another cancelable context created based on `ctx1`, so it has its own done channel, with a closure to close that done channel. It also has a goroutine that waits for either the parent done channel or the ctx2 done channel to close. If the parent done channel is closed, it cancels ctx2, and all contexts created based on ctx2. If ctx2 is canceled, the goroutine simply terminates. That's the reason why you must call the cancel function: if the context is never canceled, the goroutine leaks.

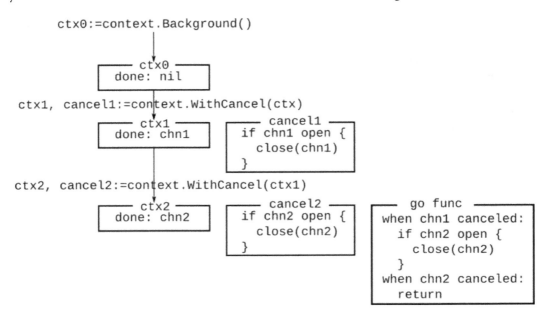

Figure 8.1 – Nested contexts and cancelation

As a side note, *Figure 8.1* gives a conceptual overview of the cancelation feature. But how can you check whether a channel is open? The actual cancel function in the standard library has quite an elaborate implementation that also includes functionality to cancel child contexts. A simple cancel function that can be called multiple times can be implemented as follows:

```
func GetCancelFunc() (cancel func(), done chan struct{}) {
    var once sync.Once
    done = make(chan struct{})
    cancel = func() {
        once.Do(func() { close(done) })
    }
    return
}
```

Context timeouts and deadlines work the same way. The only difference with a cancelable context is that a context with a deadline or timeout has a timer that will call the cancel function once the deadline passes. A timeout works with a duration:

```
ctx2, cancel := context.WithTimeout(ctx,2*time.Second)
defer cancel()
```

A deadline works with a time:

```
d := time.Now().Add(2*time.Second)
ctx2, cancel := context.WithDeadline(ctx, d)
defer cancel()
```

When a context is canceled, the `Err()` method will be set to a `context.Canceled` error. When a context times out, the `Err()` method will return a `context.DeadlineExceeded` error.

Contexts also offer a mechanism to store request-specific values. However, you should not treat this mechanism as a generic `map[any]any` storage. As I mentioned before, contexts are implemented using a decorator pattern. Every new addition to a context creates a new context with that addition while leaving the old one intact. This is also true for the values stored in a context. If you add a value to a context, you will get a new context that has that value. When you query a context for a value (`ctx.Value(key)`) and the key does not match what `ctx` has, it will call its parent to search for that value, and the call will continue recursively until the key is found. This means two things: first, you can override an existing value in a new context. The users of the new context will see the new value, whereas the users of the old context will see the unmodified value. Second, if you add hundreds of values to a context, you'll get a chain of hundreds of contexts. So, be mindful about what and how much put into a context. If you need to add a lot of values, add a single structure with many values.

In a simple program, there is nothing wrong with using strings as keys for context values. However, this is open to misuse and can cause very hard-to-diagnose subtle bugs if multiple packages use the same name to add values that mean different things. Because of this, the idiomatic way to deal with values in a context is by using the Go type system to prevent unintentional key collisions – that is, use a different type for each key. The following example illustrates adding a request identifier to the context:

```
1: type requestIDKeyType int
2: var requestIDKey requestIDKeyType
3:
4: func WithRequestID(ctx context.Context) context.Context {
5:     return context.WithValue(ctx, requestIDKey,
           uuid.New())
6: }
7:
```

```
 8: func GetRequestID(ctx context.Context) uuid.UUID {
 9:     id, _ := ctx.Value(requestIDKey).(uuid.UUID)
10:     return id
11: }
```

Line 1 defines an unexported data type, and line 2 defines a context key using this data type. This type of declaration ensures that nobody can create the same key unintentionally in a different package. Line 4 defines the `WithRequestID` function that returns a new context with added request identifier. Line 8 defines the `GetRequestID` function that extracts the request identifier from the context. If the context does not have a request identifier, it will return the zero value for UUID (which is a byte array of zeros). Based on this, can you guess what the following program will print?

```
ctx  := context.Background()
ctx1 := WithRequestID(ctx)
ctx2 := WithRequestID(ctx1)
fmt.Println(GetRequestID(ctx), GetRequestID(ctx1),
GetRequestID(ctx2))
```

It will actually print a different output every time you run it. However, the first output will always be 00000000-0000-0000-0000-000000000000 (the zero value for UUID), the second value will be the request identifier in `ctx1`, and the third value will be the request identifier in `ctx2`. Note that adding another request identifier to the context does not overwrite the request identifier for `ctx1`.

A common question is: what values should be put in a `Context`? The guiding principle is whether or not the value is request-specific rather than the nature of the value itself. If you have a database connection that is shared among all request handers, that does not belong in the context. However, if you have a system that may connect to a different database based on the caller's credentials, then it may make sense to put that database connection into the context. An application configuration structure does not belong in the context. If you load a configuration item from a database based on the request, it may make sense to put that into the context.

A context object is meant to be passed into multiple goroutines, which means you have to be careful about race conditions involving context values. Consider the following context:

```
newCtx:=context.WithValue(ctx,mapKey,map[string]
interface{"key":"value"})
```

If newCtx is passed to multiple goroutines, the map in that context becomes a shared variable. Multiple goroutines adding/removing values to/from that map will cause race conditions and probably corrupt memory. A correct way of dealing with this problem is using a structure:

```
type StructWithMap struct {
    sync.Mutex
```

```
    M map[string]interface{}
}
...
newCtx:=context.WithValue(ctx,mapKey,&StructWithMap{
    M:make(map[string]interface{}),
}
```

In this example, a pointer to a structure with a mutex and map is put into the context. Goroutines will have to lock the mutex to access the map. Also, note that a mutex must not be copied, so the address of the structure is put into the context.

Backend services

If you are using Go, it is likely that you have written or will write a backend service of some sort. Service development comes with a unique set of challenges. First, the concurrency aspect of request handling is usually hidden under a service framework, which causes unintentional memory sharing and data races. Second, not all clients of the service have good intentions (attacks) or are free of bugs. In this section, we will look at some basic constructs using HTTP and web socket services. But before those, knowing a bit about TCP networking helps because many higher-level protocols like HTTP and web sockets are based on TCP. Next, we will construct a simple TCP server that handles requests concurrently and shuts down gracefully. For this, we need a listener, a request handler, and a wait group:

```
type TCPServer struct {
    Listener    net.Listener
    HandlerFunc func(context.Context,net.Conn)
    wg sync.WaitGroup
}
```

The server provides a Listen method that waits for connections. It starts by creating a cancelable context. This context will be canceled when the Listen method returns, notifying all active connection handlers about the cancelation. When a connection is established, the method creates a new goroutine to handle the connection and continues listening:

```
func (srv *TCPServer) Listen() error {
    baseContext, cancel :=
        context.WithCancel(context.Background())
    defer cancel()
    for {
        conn, err := srv.Listener.Accept()
        if err != nil {
```

```go
            if errors.Is(err, net.ErrClosed) {
                return nil
            }
            fmt.Println(err)
        }
        srv.wg.Add(1)
        go func() {
            defer srv.wg.Done()
            srv.HandlerFunc(baseContext, conn)
        }()
    }
}
```

As you may notice, the `Listen` method will not return until the `Accept` call fails. Once started, the server can be stopped from another goroutine by closing the listener:

```go
func (srv *TCPServer) StopListener() error {
    return srv.Listener.Close()
}
```

Closing the listener will cause the `Accept` call to fail, and the `Listen` method will stop listening, cancel the context, and return. Canceling the context will notify all active connections that the server is shutting down, but it is unreasonable to expect them to stop processing immediately. We have to give some time to these handlers to finish, using a `WaitForConnections` method with a timeout:

```go
func (srv *TCPServer) WaitForConnections(timeout time.Duration) {
    toCh := time.After(timeout)
    doneCh := make(chan struct{})
    go func() {
        srv.wg.Wait()
        close(doneCh)
    }()
    select {
    case <-toCh:
    case <-doneCh:
    }
}
```

This is where `WaitGroup` is useful. If there are no active connections, `srv.wg.Wait()` will immediately return, closing `doneCh`, which will cause the `WaitForConnections` method to return. If there are active connections, we will wait for them in a separate goroutine, and if they all complete before the timeout, `doneCh` will be closed and the method will return. However, if there are connections that do not comply with the stop request within the given timeout, the method will still return, leaving those connections active. An option to deal with this is to close those active connections, but that may result in unexpected behavior. So, it will be up to you to decide the best course of action in this situation.

Containerized backend services can handle termination signals for a graceful shutdown. This has to be done before any of the servers start listening. The following code snippet will set up a signal handle to listen to termination signals, and will give the server 5 seconds to shutdown:

```
sig := make(chan os.Signal, 1)
signal.Notify(sig, syscall.SIGTERM, syscall.SIGINT)
go func() {
      <-sig
      srv.StopListener()
      srv.WaitForConnections(5 * time.Second)
}()
```

The following is a simple echo server using these utilities:

```
srv.Listener, err = net.Listen("tcp", "")
if err != nil {
      panic(err)
}
srv.HandlerFunc = func(ctx context.Context, conn net.Conn) {
      defer conn.Close()
      // Echo server
      io.Copy(conn, conn)
}
srv.Listen()
srv.WaitForConnections(5 * time.Second)
```

Now let's look at a simple HTTP service using the standard library. HTTP is a text-based protocol built on top of TCP, so the HTTP code very much looks like the TCP server. An HTTP request contains a header that tells the HTTP verb (GET, POST, etc.), the path (part of the URL after the host and the port), and HTTP headers. Usually, an HTTP server uses different handlers for different request types, and which handler to call is determined based on the request header information (HTTP verb, path, and headers). This is called **request multiplexing**. The Go standard library includes a basic multiplexer; there are many frameworks out there that offer different capabilities with varying performance characteristics. But when it comes to the application level where the requests are handled, there are a few key assumptions you have to keep in mind:

- Request handlers will be called concurrently.

- Requests may be received out of order. That is, the client calling the APIs in a certain order does not mean that the server will receive those calls in the same order.

- You cannot trust the caller. You must authenticate the caller for APIs that provide access to privileged resources, limit the size of data the caller can send or receive, and validate the data your API receives.

The standard library provides an `http.ServeMux` type that can be used as a simple request multiplexer. You can register request handlers to an instance of `http.ServeMux` using `Handler` and `HandlerFunc` methods. The standard library also declares a default `ServeMux` instance, so the `http.Handler` and `http.HandlerFunc` functions can be used to register request handlers to that particular instance. So, you can do the following:

```
mux := http.NewServeMux()
svc := NewDashboardService()
mux.HandleFunc("/dashboard/", svc.DashboardHandler)
http.ListenAndServe("localhost:10001", mux)
```

What's happening here is that we create a multiplexer, create an instance of our backend service implementation (which is a hypothetical dashboard service), register the handler for the `/dashboard/` path, and start the server. The rest of the request handling happened in the `DashboardHandler` method. The Go type system allows passing a method for function variables, so in this case, the request handler is a method that has access to the `DashBoardService` implementation, so this can contain all the configuration information, database connections, clients for remote services, and so on. An important point to note here is that the request handler methods will be called concurrently, and all of them will be using the `svc` instance declared previously. Thus, if you need to modify anything in `svc`, you have to protect it with a mutex.

A common pattern supported by many multiplexers is the building of a call chain using middleware functions. A middleware in this context is a function that performs some operation on the request such as authentication or context preparation and passes it on to the next handler in the chain. The next handler can be the actual request handler or another middleware. For example, the following middleware replaces the request body with a limited reader to protect the service from large requests:

```
func Limit(maxSize int64, next http.HandlerFunc) http.
HandlerFunc {
    return http.HandlerFunc(func(w http.ResponseWriter,
      req *http.Request) {
```

```
        req.Body = http.MaxBytesReader(w, req.Body,
          maxSize)
        next(w, req)
    })
}
```

And the following middleware authenticates the caller using a given authenticator function and adds the user identifier to the context. Note that if authentication fails, the next handler is not even called:

```
func Authenticate(authenticator func(*http.Request) (string,
error), next http.HandlerFunc) http.HandlerFunc {
    return http.HandlerFunc(func(w http.ResponseWriter,
      req *http.Request) {
        userId, err := authenticate(req)
        if err != nil {
            http.Error(w, err.Error(),
              http.StatusUnauthorized)
        return
    }
    next(w, req.WithContext(WithUserID(req.Context(),
        userId)))
    })
}
```

The handler registration now becomes this:

```
mux.HandleFunc("/dashboard/", Authenticate(authFunc,
  Limit(10240, svc.DashboardHandler)))
```

After this setup, DashboardHandler is guaranteed to receive authenticated requests that do not exceed 10 Kb in size.

Next, let's look at the handler itself. This handler responds to a GET request by computing and returning some dashboard data, which is composed of summary information from multiple backend services. The POST request is used to set dashboard parameters for a user. So, the handler looks like this:

```
func (svc *DashboardService) DashboardHandler(w http.
ResponseWriter, req *http.Request) {
    switch req.Method {
    case http.MethodGet:
```

```
                dashboard := svc.GetDashboardData(req.Context(),
                    GetUserID(req.Context())
                json.NewEncoder(w).Encode(dashboard)
        case http.MethodPost:
                var params DashboardParams
                if err := json.NewDecoder(req.Body)
                    .Decode(&params); err != nil {
                        http.Error(w, err.Error(),
                            http.StatusBadRequest)
                }
                svc.SetDashboardConfig(req.Context(),
                    GetUserID(req.Context()), params)
        default:
                http.Error(w, "Unhandled request type",
                    http.StatusMethodNotAllowed)
        }
}
```

As you can see, this code relies on middleware for authentication and limiting request size. Let's look at the `GetDashboardData` method.

Distributing work and collecting results

Our hypothetical server talks to two backend services to collect statistics. The first service returns information about the current user, and the second returns information about the account, which may include multiple users. In this example, we modeled these as some opaque backend services, but in reality, these can be other web services, microservices called via gRPC, or database calls:

```
type DashboardService struct {
    Users    UserSvc
    Accounts AccountSvc
}

type DashboardData struct {
    UserData     UserStats
    AccountData  AccountStats
    LastTransactions []Transaction
}
```

The actual handler illustrates several methods of distributing work to multiple goroutines and collecting results from them:

```
func (svc *DashboardService) GetDashboardData(ctx context.
Context, userID string) DashboardData {
result := DashboardData{}
wg := sync.WaitGroup{}
```

1. The first goroutine calls the `Users` service to collect statistics for the given user identifier. It uses a wait group to notify the completion of the work and directly modifies the result structure. This is safe as long as no other goroutine touches the `result.UserData` field. If the context is canceled, it will be up to the `Users.GetStats` method to return as soon as possible:

    ```
    wg.Add(1)
    go func() {
        defer wg.Done()
        var err error
        // Setting result.UserData is safe here, because
        // this is the only gouroutine accessing
        // that field
        result.UserData, err = svc.Users.GetStats(ctx,
            userID)
        if err != nil {
            log.Println(err)
        }
    } ()
    ```

2. The second goroutine gets the account level statistics via a channel, but with a timeout of 100 milliseconds. That means the `Accounts.GetStats()` method creates a goroutine to compute the statistics and returns it asynchronously. When this result is read, it is sent to the `acctCh` channel in the `select` statement. The `select` statement also detects context cancelation. If the context is canceled while the `Accounts.GetStats` method is running, it may continue running after the handler returned, but it should eventually realize the context is canceled and return. If the context is canceled because of a timeout, the zero value for the account data will be returned:

    ```
    acctCh := make(chan AccountStats)
    go func() {
        // Make sure acctCh is closed when goroutine
        // returns, so we don't indefinitely block
    ```

```
          // waiting for a result from it
          defer close(acctCh)
          newCtx, cancel := context.WithTimeout(ctx,
             100*time.Millisecond)
          defer cancel()
          resultCh := svc.Accounts.GetStats(newCtx, userID)
          select {
          case data := <-resultCh:
                acctCh <- data
          case <-newCtx.Done():
          }
    }()
```

3.   The third part creates two goroutines (one for users and the other for accounts) that collect transaction information. These goroutines write the transaction information asynchronously to a common channel, which is listened to by another goroutine that fills the LastTransactions slice. There is a separate wait group that is waited in a new goroutine that closes the transaction channel once all data elements are received:

```
transactionWg := sync.WaitGroup{}
transactionWg.Add(2)
transactionCh := make(chan Transaction)
go func() {
      defer transactionWg.Done()
      for t := range svc.Users.GetTransactions(ctx,
         userID) {
            transactionCh <- t
      }
}()
go func() {
      defer transactionWg.Done()
      for t := range svc.Accounts.GetTransactions(ctx,
         userID) {
            transactionCh <- t
      }
}()
go func() {
      transactionWg.Wait()
```

```
            close(transactionCh)
    }()
```

4. The next goroutine collects transactions from `transactionCh`. Note that this is a fan-in operation:

```
wg.Add(1)
go func() {
    defer wg.Done()
    for record := range transactionCh {
        // Updating result.LastTransactions is
        // safe here because this is the
        // only goroutine that sets it
        result.LastTransactions =
        append(result.LastTransactions, record)
    }
}()
```

5. As a final step, we wait for all the goroutines to complete, read the account data from its channel, and return. The receive from `acctCh` will not block indefinitely because it either returns a value or it is closed, in which case, we return the zero value for `AccountData`:

```
wg.Wait()result.AccountData = <-acctCh
return result
}
```

This example demonstrates several methods for distributing work and collecting results: one using shared memory safely, and the others using channels. If you are using shared memory, take extra care to protect variables accessed by multiple goroutines. If you are using channel communications, make sure all goroutines terminate correctly.

Semaphores – limiting concurrency

What happens if you want to limit concurrency? The dashboard handler can be quite expensive, and you might want to limit the number of concurrent calls to it. A semaphore can be used for this purpose. Semaphores are versatile concurrency primitives. A semaphore keeps a counter representing the number of resources available. The term "resource" should be taken in an abstract sense: it can refer to an actual computing resource, or it can simply mean permission to enter a critical section. A thread consumes resources by decrementing the value of the counter and relinquishes them by incrementing the counter. If the counter is zero, consuming the resource is not allowed, and the thread blocks until the counter is non-zero again. So, a semaphore is like a mutex with a counter. Or, to put it in another way, a mutex is a binary semaphore. You can use a channel of capacity N as a semaphore to control access to N instances of a resource:

```
semaphore := make(chan struct{},N)
```

You can acquire a resource with a send operation. This operation will block if the semaphore buffer is full:

```
semaphore <- struct{}{}
```

You can relinquish the resource with a receive operation. This will wake up other goroutines waiting to acquire the resource:

```
<- semaphore
```

We are using a channel of the `struct{}` type here (whose size is 0), so the channel buffer does not actually use any memory.

This is a good way of limiting concurrency in a program where a potentially unbounded number of goroutines can be created. The following example shows a middleware that limits concurrent calls to the dashboard handler:

```
func ConcurrencyLimiter(sem chan struct{}, next http.
HandlerFunc) http.HandlerFunc {
        return http.HandlerFunc(func(w http.ResponseWriter,
        req *http.Request) {
            sem <- struct{}{}
            defer func() { <-sem }()
            next(w, req)
    })
}
```

And a concurrency-limited handler can be defined using the following:

```
mux.HandleFunc("/dashboard/", ConcurrencyLimiter(make(chan
struct{}, limit), svc.DashboardHandler))
```

Streaming data

A typical software engineer's life revolves around moving and transforming data. Sometimes the data being moved or transformed does not have a predefined size limit, or it is produced in a piecemeal fashion, so it is not reasonable to load it all and process it. That's when you may need to stream data.

When I say *streaming*, what I mean is the processing of data generated continuously. This includes dealing with an actual stream of bytes, such as transferring a large file, as well as dealing with a list of structured objects such as records retrieved from the database, or time-series data generated by sensors. So, you usually need a "generator" function that will collect data based on a specification and pass it on to the subsequent layers.

In what follows, we will build a (hypothetical) application that deals with time series data stored in a database. The application will use a query to select a subset of the data in the database, compute a running average, and return the instances when the running average goes above a certain threshold.

First, the generator: the following declares a `Store` data type that contains the database information in it. An instance of `Store` will be initialized at program startup with a connection to the database:

```
type Store struct {
    DB *sql.DB // The database connection
}
```

The `Entry` structure contains a measurement performed at a certain time:

```
type Entry struct {
    At    time.Time
    Value float64
    Error error
}
```

Why is there an `Error` in the `Entry` structure? Error reporting and handling is one of the important considerations of streaming results because errors can happen at every stage of streaming: during preparation (when you run the query, for instance), during the actual streaming (retrieval of some of the entries can fail), and after all the elements are processed (did the stream stop because everything is sent, or because something unexpected happened?). Unlike synchronous processing scenarios that result in either data or an error, a stream can include multiple errors as well as data elements. So, it is best to pass these errors along with each entry so that the processing logic downstream can decide how to deal with errors.

The following illustrates the general structure of such generator methods. It is designed as a method of `Store` so it has access to database connection information. The method gets a context together with a query structure that describes what results are requested. It returns a channel of the `Entry` type from which the caller can receive query results and an error that describes an error that happened at the preparation stage (for instance, a query error):

```
func (svc Store) Stream(ctx context.Context, req Request) (<-
chan Entry, error) {
        // Normally you should build a query using
```

```go
    // the request
    rows, err := svc.DB.Query(`select at,
      value from measurements`)
    if err != nil {
        return nil, err
    }
    ret := make(chan Entry)
    go func() {
        // Close the channel to notify the receiver
        // that data stream is finished
        defer close(ret)
        // Close the database result set
        defer rows.Close()
        for {
            var at int64
            var entry Entry
            // Check for cancelations
            select {
            case <-ctx.Done():
                return
            default:
            }
            if !rows.Next() {
                break
            }
            if err := rows.Scan(&at,
              &entry.Value); err != nil {
                ret <- Entry{Error: err}
                continue
            }
            entry.At = time.UnixMilli(at)
            ret <- entry
        }
        if err := rows.Err(); err != nil {
            ret <- Entry{Error: err}
        }
```

```
        }()
    return ret, nil
}
```

The method prepares a database query based on the request and runs it. Any errors at this stage are returned immediately as an `error` value from the method. If the query runs successfully, the method starts a goroutine that will retrieve results from the database, and it returns a channel from which the caller can read the results one by one. At any point, the caller can cancel the generator method by canceling the context. The goroutine starts by deferring some cleanup tasks, namely, the closing of the database result set and the closing of the results channel. The goroutine will iterate through the result set and send the results via the channel one by one. Any errors captured while iterating the results will be sent in that instance of the `Entry` structure. When all data items are sent, the goroutine will close the channel, signaling the exhaustion of the results. If the result set fails, an additional `Entry` instance will be sent with the error.

What is really happening here is that the `Stream` method creates a closure that sends data through a channel. That means the closure will live after the `Stream` method returns. Thus, any cleanup that needs to be done is done in the closure, not in the `Stream` method itself. It is also important to ensure termination of the closure either by consuming all the results or by canceling the context; otherwise, the goroutine (and the database resources associated with it) will leak.

Stream processing is structurally similar to data pipelines. Stream processing components can be chained one after the other to process data in an efficient manner. For example, the following function reads the input stream and filters out entries that are below a certain value while preserving error entries:

```
func MinFilter(min float64, in chan<- store.Entry) <-chan
store.Entry {
    outCh := make(chan store.Entry)
    go func() {
        defer close(outCh)
        for entry := range in {
            if entry.Err != nil ||
                entry.Value >= min {
                outCh <- entry
            }
        }
    }()
    return outCh
}
```

Sometimes you need to separate the stream into multiple streams based on certain criteria. The following function returns a closure that sends all errors to a separate channel. The returned entry channel now only contains the entries that have no error:

```go
func ErrFilter(in <-chan store.Entry) (<-chan store.Entry,
<-chan error) {
    outCh := make(chan store.Entry)
    errCh := make(chan error)
    go func() {
        defer close(outCh)
        defer close(errCh)
        for entry := range in {
            if entry.Error != nil {
                errCh <- entry.Error
            } else {
                outCh <- entry
            }
        }
    }()
    return outCh, errCh
}
```

After filtering the stream and separating out the errors, we can compute a moving average of the measurements and select the entries when the moving average is above a threshold. For this, we define the following new structure, which contains the entry and the moving average value:

```go
type AboveThresholdEntry struct {
    store.Entry
    Avg float64
}
```

The following function then reads the entries from the input channel and keeps a moving average of the measurements. A moving average is defined by the average value of the last windowSize elements seen in the stream. When a new measurement is read, the first measurement is removed from the running total, and the new measurement is added to it. This requires a *first-in, first-out*, or a circular buffer of the given size. A channel can double as such a buffer:

```go
func MovingAvg(threshold float64, windowSize int, in <-chan
store.Entry) <-chan AboveThresholdEntry {
    // A channel can be used as a circular/FIFO buffer
```

```
        window := make(chan float64, windowSize)
        out := make(chan AboveThresholdEntry)
        go func() {
                defer close(out)
                var runningTotal float64
                for input := range in {
                        if len(window) == windowSize {
                                avg := runningTotal /
                                        float64(windowSize)
                                if avg > threshold {
                                        out <- AboveThresholdEntry{
                                        Entry: input,
                                        Avg:   avg,
                                        }
                                }
                                // Drop the oldest value in window
                                runningTotal -= <-window
                        }
                        // Add value to window
                        window <- input
                        runningTotal += input
                }
        }()
        return out
}
```

The following code snippet puts all of it together. It streams the results from the database, filters the results, computes a moving average, and writes the selected entries to the output. If there were any errors during this processing, it writes the first error after all the output is written:

```
// Stream results
ctx, cancel := context.WithCancel(context.Background())
defer cancel()
entries, err := st.Stream(ctx, store.Request{})
if err != nil {
        panic(err)
}
```

```go
// Remove all entries less than 0.001
filteredEntries := filters.MinFilter(0.001, entries)
// Split errors
entryCh, errCh := filters.ErrFilter(filteredEntries)
// Select all entries when moving average >0.5 with
// window size of 5
resultCh := filters.MovingAvg(0.5, 5, entryCh)
// We will capture the first error, and cancel
var streamErr error
go func() {
    for err := range errCh {
    // Capture first error
        if streamErr == nil {
            streamErr = err
            cancel()
        }
    }
}()
for entry := range resultCh {
    fmt.Printf("%+v\n", entry)
}
if streamErr != nil {
    fmt.Println(streamErr)
}
```

There are a few points to be careful about here. First, there is a separate goroutine that is receiving from the `error` channel. When the first error is captured, it cancels the stream processing completely. If that happens, the `Stream` method receives that cancelation and will close the entries channel. This will be detected by the next processing step in the pipeline (`MinFilter`), and it will close its channel. This will continue until `resultCh` is closed, and when that happens, the `for` loop that is reading from `resultCh` will close as well. The next statement reads the `streamErr` variable, which is written in the error handling goroutine, but this is not a data race. The `ErrFilter` function closes `errCh` before it closes `entryCh`, and `entryCh` is closed before `resultCh` (can you see why?), thus the termination of the `for` loop guarantees that `errCh` is closed. Second, the results are collected in the main goroutine. The same result can also be achieved using a separate goroutine to collect the results, but then you have to use `sync.WaitGroup` to wait for both goroutines to finish. You can also choose to read the errors in the main goroutine while collecting the results in a different goroutine. There, you have to use `sync.WaitGroup` again, because the closing of `errCh` happens before the closing of `resultCh`, so you have to wait for `resultCh` to close.

Not all data streaming implementations can be chained using Go concurrency primitives like this. If, for instance, you have a microservice architecture that uses HTTP requests, a WebSocket, or a remote procedure call scheme such as gRPC, then you can't really chain components using channels. Some of these components will be located on different nodes on a network so communication between them will be through a network connection. However, the basic constructs we discussed previously can still be used with the help of simple adapters. So let's take a look at how such adapters can be implemented to utilize Go concurrency primitives effectively. First, we need to decide what our objects look like when they are exchanged between different components on a network. So we need to serialize (or marshal) these data objects and send them over the wire, where they can be deserialized (or unmarshaled) to reconstruct the original object, or something as close to that as possible. Using an RPC implementation such as gRPC helps greatly in these situations by forcing you to think and model your objects using marshalable/unmarshalable objects only. However, that is not always the case. A common format for data exchange is JSON, so we will use JSON in this example. You can immediately realize the potential problem here: the store.Entry structure can be marshaled easily, but when unmarshaled, the Entry.Error field cannot be reconstructed. If you are sending errors over a network connection, you should implement error structures containing both type and diagnostic information so they can be reconstructed properly on the receiving end. For the sake of simplicity, we will simply represent errors as strings:

```
type Message struct {
    At      time.Time `json:"at"`
    Value float64     `json:"value"`
    Error string      `json:"err"`
}
```

Here, the Message structure is a serializable version of the store.Entry type. When sending store.Entry type objects over a network connection, we first translate each entry to a message, encode it as JSON, and write it. Since we are dealing with streaming multiple such store.Entry structures, we have a channel from which we read the stream. A simple generic adapter that does this is as follows:

```
func EncodeFromChan[T any](input <-chan T, encode func(T) ([]
byte, error), out io.Writer) <-chan error {
    ret := make(chan error, 1)
    go func() {
        defer close(ret)
        for entry := range input {
            data, err := encode(entry)
            if err != nil {
                ret <- err
                return
```

```
        }
        if _, err := out.Write(data); err != nil {
            if !errors.Is(err, io.EOF) {
                ret <- err
            }
            return
        }
    }
}()
return ret
}
```

This function reads entries from the given channel, serializes them using the given `encode` function, and writes the result to the given `io.Writer`. Note that it returns an `error` channel, which has a capacity of 1. The channel passes error information to the caller, and since it has a capacity of 1, the error can be sent to the channel without blocking, even if the caller is not receiving from that channel. The same channel also serves as a signal for completion. An HTTP handler using this is as follows:

```
http.HandleFunc("/db", func(w http.ResponseWriter, req *http.
Request) {
    storeRequest := parseRequest(req)
    data, err := st.Stream(req.Context(), storeRequest)
    if err != nil {
        http.Error(w, "Store
            error", http.StatusInternalServerError)
        return
    }
    errCh := EncodeFromChan(data, func(entry store.Entry)
      ([]byte, error) {
        msg := Message{
            At:    entry.At,
            Value: entry.Value,
        }
        if entry.Error != nil {
            msg.Error = entry.Error.Error()
        }
        return json.Marshal(msg)
    }, w)
```

```
        err = <-errCh
        if err != nil {
            fmt.Println("Encode error", err)
        }
    }))
}))
```

There are several points to note here. The handler calls the `store.Stream` function using the request context. Because of this, if the caller of this API stops listening for the stream and closes the connection, the context will be canceled and the handler will stop generating results. Second, the error from the store can be returned as an HTTP error, but not the error from the encoder. This is because by the time an error is detected in the streaming, the HTTP response header will already be written with a `200 Ok` HTTP status, so there is no way to change it. The best thing that can be done is to stop processing and log the error. Note that this situation does not include the entries retrieved from the store with an error. Those are transferred successfully. The error only happens if marshaling fails, or if writing to the network connection fails, which can happen if the caller terminates the connection.

Similar to the encoding function, we need a decoder for the receiving end of the connection. The following generic function reads and decodes messages, and sends them over a channel. The actual reading from the connection is to be implemented in the given decode function:

```
func DecodeToChan[T any](decode func(*T) error) (<-chan T,
<-chan error) {
    ret := make(chan T)
    errch := make(chan error, 1)
    go func() {
        defer close(ret)
        defer close(errch)
        var entry T
        for {
            if err := decode(&entry); err != nil {
                if !errors.Is(err, io.EOF) {
                    errch <- err
                }
                return
            }
            ret <- entry
        }
    }()
    return ret, errch
}
```

There are two channels returned this time: one for the actual data stream and one for the error. Again, the error channel has a capacity of 1, so the caller does not need to listen to it. An HTTP client that calls this API and streams data is as follows:

```
resp, err := http.Get(APIAddr+"/db")
if err != nil {
    panic(err)
}
defer resp.Body.Close()
decoder := json.NewDecoder(resp.Body)
entries, rcvErr := DecodeToChan[store.Entry](
  func(entry *store.Entry) error {
    var msg Message
    if err := decoder.Decode(&msg); err != nil {
        return err
    }
    entry.At = msg.At
    entry.Value = msg.Value
    if msg.Error != "" {
        entry.Error = fmt.Errorf(msg.Error)
    }
    return nil
})
```

As you can see, this is a straightforward HTTP call that uses a JSON decoder to decode a stream of Message objects from the response and sends them to the entries channel. Now, this channel can be fed to the stream processing pipeline. A separate goroutine can be used to listen to errors from the error channel.

This example illustrates how you can translate between readers/writers and channels when dealing with streams. Streaming results like this uses very little memory, starts returning results quickly, and scales well because data will be processed as it arrives. The next example will use WebSockets to illustrate how you can deal with multiple streams concurrently.

Dealing with multiple streams

Many times, you have to coordinate between data coming from and going to multiple streams concurrently. A simple example would be a chat room server using WebSockets. Unlike standard HTTP, which is composed of request/response pairs, WebSockets use bidirectional communication over HTTP, so you can both read from and write to the same connection. They are ideal for long-running conversations between systems where both sides send and receive data, such as this chat room example. We will develop a chat room server that accepts WebSocket connections from multiple clients. The server will distribute a message it receives from a client to all the clients connected at that moment. For this purpose, we define the following message structure:

```
type Message struct {
    Timestamp time.Time
    Message   string
    From      string
}
```

Let's start with the client. Each client will connect to the chat server using a WebSocket:

```
cli, err := websocket.Dial("ws://"+os.Args[1]+"/chat", "",
"http://"+os.Args[1])
if err != nil {
    panic(err)
}
defer cli.Close()
```

Clients will read text input from the terminal, and send it to the chat server through that WebSocket. At the same time, all clients will be listening to incoming messages as well. So it is clear that we need to have several goroutines to make these things concurrently. We start with setting up the channels to send and receive messages to and from the server. In the following code, rcvCh will be used to receive messages received from the server, and inputCh will be used to send messages to the server:

```
decoder := json.NewDecoder(cli)
rcvCh, rcvErrCh := chat.DecodeToChan(func(msg *chat.Message)
error {
    return decoder.Decode(msg)
})
sendCh := make(chan chat.Message)
sendErrCh := chat.EncodeFromChan(sendCh, func(msg chat.Message)
([]byte, error) {
    return json.Marshal(msg)
}, cli)
```

Next, using a separate goroutine, we read text from the terminal and send it to the server:

```
done := make(chan struct{})
go func() {
    scanner := bufio.NewScanner(os.Stdin)
```

```go
    for scanner.Scan() {
        text := scanner.Text()
        select {
        case <-done:
            return
        default:
        }
        sendCh <- chat.Message{
            Message: text,
        }
    }
}()
```

The final piece of the client code deals with the messages received from the server:

```go
for {
    select {
    case msg, ok := <-rcvCh:
        if !ok {
            close(done)
            return
        }
        fmt.Println(msg)
    case <-sendErrCh:
        return
    case <-rcvErrCh:
        return
    }
}
```

The server is a bit more involved as it has to distribute messages received from the clients to all the connected clients. It also has to keep track of the connected clients and make sure a malicious client cannot disrupt the whole system. The server will have a `handler` function containing the decoder and encoder goroutines, similar to the ones we have for the client. However, there are some significant differences. First, the server creates a separate goroutine for each connected client. That means if we need to keep track of all active connections, we need a share data structure, and thus a mutex to protect. But there is a way to do this without any shared memory (and thus, without any risk of memory races.) Instead of a shared memory structure, we create a controller goroutine that will keep track of all active connections, and dispatch any received message to them. When a new connection is established, we will use a channel, connectCh, to send the data channel for that connection. When the connection is closed, we will use a different channel, disconnectCh, to send a notification that a disconnect has happened. We will also use a data channel that will receive the messages:

```
dispatch := make(chan chat.Message)
connectCh := make(chan chan chat.Message)
disconnectCh := make(chan chan chat.Message)
go func() {
    clients := make(map[chan chat.Message]struct{})
    for {
        select {
        case c := <-connectCh:
            clients[c] = struct{}{}
        case c := <-disconnectCh:
            delete(clients, c)
        case msg := <-dispatch:
            for c := range clients {
                select {
                case c <- msg:
                default:
                    delete(clients, c)
                    close(c)
                }
            }
        }
    }
}()
```

The connection handler deals with the actual encoding and decoding of data:

```
http.Handle("/chat", websocket.Handler(func(conn *websocket.
Conn) {
    client := conn.RemoteAddr().String()
    inputCh := make(chan chat.Message, 10)
    connectCh <- inputCh
    defer func() {
        disconnectCh <- inputCh
    }()
```

```go
        decoder := json.NewDecoder(conn)
        data, decodeErrCh := chat.DecodeToChan(func(msg
          *chat.Message) error {
            err := decoder.Decode(msg)
            msg.From = client
            msg.Timestamp = time.Now()
            return err
        })
        encodeErrCh := chat.EncodeFromChan(inputCh, func(msg
          chat.Message) ([]byte, error) {
            return json.Marshal(msg)
        }, conn)
        for {
            select {
            case msg, ok := <-data:
                if !ok {
                    return
                }
                dispatch <- msg
            case <-decodeErrCh:
                return
            case <-encodeErrCh:
                return
            }
        }
    }))
```

As you can see, when a new connection starts, an `input` channel is constructed to accept messages coming from all the clients. This is a buffered channel to prevent a malicious client that stops reading from the WebSocket without closing it. The `input` channel will buffer the last 10 messages, and if these messages cannot be sent, the controller will terminate the connection for that client by closing the `data` channel. When the `data` channel is closed, the encoding goroutine will terminate, eventually terminating the handler for that client.

This simple server illustrates a way to distribute data streams across multiple clients without falling into memory race issues. Many algorithms that look like they need a shared data structure can be converted into a message-passing algorithm that doesn't need one. This is not always possible, so you should evaluate both ways when developing such programs. If you are writing a cache, then shared memory with mutexes makes more sense. If you are coordinating work between multiple goroutines, a separate goroutine using multiple channels makes more sense. Use your judgment, try writing it, and if you end up with spaghetti code, throw it away and use a different approach.

Summary

This chapter was about the language utilities and concurrency patterns for dealing with requests – mainly, requests that come through the network. In an evolving architecture, it is often the case that some components developed for a non-networked system do not perform as expected when applications move to a more service-oriented architecture. I hope that knowing the basic principles and design rationale behind these utilities and patterns will help you when you are faced with these problems.

Next, we will look at atomics, why you should be careful when you are using them, and how they can be used effectively.

9

Atomic Memory Operations

Atomic memory operations provide the low-level foundation necessary to implement other synchronization primitives. In general, you can replace all atomic operations of a concurrent algorithm with mutexes and channels. Nevertheless, they are interesting and sometimes confusing constructs, and you should know how they work. If you use them carefully, they can become good tools for code optimization without increasing complexity.

In this chapter, we will explore the following topics:

- Memory guarantees of atomic memory operations
- The compare-and-swap operation
- Practical uses of atomics, including counters, heartbeats/progress meters, cancellations, and detecting change

Technical Requirements

The source code for this particular chapter is available on GitHub at `https://github.com/PacktPublishing/Effective-Concurrency-in-Go/tree/main/chapter9`.

Memory guarantees

Why do we need separate functions for atomic memory operations? If we write to a variable whose size is less or equal to the machine word size (which is what the `int` type is defined to be), such as `a=1`, wouldn't that be atomic? The Go memory model actually guarantees that the write operation will be atomic; however, it does not guarantee when other goroutines will see the effects of that write operation, if ever. Let's try to dissect what this statement means. The first part simply says that if you write to a shared memory location that is the same size as a machine word (i.e., `int`) from one goroutine and read it from another, you will not observe some random value even if there is a race. The memory model guarantees that you will only observe the value before the write operation, or the value after it (this is not true for all languages.) This also means that if the write operation is larger than the machine word size, then a goroutine reading this value may see the underlying object in an inconsistent state. For example, a `string` value includes two values, a pointer to the underlying array, and the string length. The write operations to these individual fields would be atomic, but a racy read may read a string with a nil array but nonzero length. The second part of the statement says that the compiler can optimize or reorder code, or that the hardware executes the memory operations out of order, in such a way that another goroutine cannot see the effects of the write operation at the expected time. The standard example that illustrates this point is the following memory race:

```
func main() {
    var str string
    var done bool
    go func() {
        str = "Done!"
        done = true
    }()
    for !done {
    }
    fmt.Println(str)
}
```

There is a memory race, because the `str` and `done` variables are written in a goroutine and read in another one without explicit synchronization. There are several ways this program can behave:

- It can print Done!.

- It can print an empty string. This means that the memory write to `done`, but not the memory write to `str`, is seen by the main goroutine.

- The program may hang. This means that the memory write to `done` is not seen by the main goroutine.

This is where atomics make a difference. The following program is race-free:

```
func main() {
    var str done atomic.Value
    var done atomic.Bool
    str.Store("")
    go func() {
        str.Store("Done!")
        done.Store(true)
    }()
    for !done.Load() {
```

```
    }
        fmt.Println(str.Load())
}
```

The memory guarantee for atomic operations is as follows. If the effect of an atomic memory write is observed by an atomic read, then the atomic write happened before the atomic read. This also guarantees that the following program will either print 1 or nothing (it will never print 0):

```
func main() {
        var done atomic.Bool
        var a int
        go func() {
                a = 1
                done.Store(true)
        }()
        if done.Load() {
                fmt.Println(a)
        }
}
```

Note that there is still a race condition here, but not a memory race. Depending on the execution order of the statements, the main goroutine may or may not see done as true. However, if the main goroutine sees done as true, then it is guaranteed that a=1.

This is one of the reasons why working with atomics can get complicated – memory ordering guarantees are conditional. They never block the running goroutine, so the fact that you tested whether an atomic read returned a certain value for a variable does not mean that it still has the same value when the body of that if statement is running. That's why you need to be careful when you are using atomics. It is easy to fall into race conditions using them, as with the previous program. Remember this – you can always write the same program without atomics.

Compare and swap

Any time you test for a condition and act based on the result, you can create a race condition. For example, the following function does not prevent mutual exclusion, despite the use of atomics:

```
var locked sync.Bool

func wrongCriticalSectionExample() {
        if !locked.Load() {
```

```
        // Another goroutine may lock it now!
        locked.Store(true)
        defer locked.Store(false)
        // This goroutine enters critical section
        // but so can another goroutine
    }
}
```

The function starts by testing whether the atomic `locked` value is `false`. Two goroutines can execute this statement simultaneously, and seeing that it is `false`, both can enter the critical section and set `locked` to `true`. What is needed here is an atomic operation encompassing both the comparison and store operations. That is the **compare-and-swap** (**CAS**) operation, which does exactly what its name implies – it compares whether a variable has the expected value and, if so, swaps that value with a given value atomically. If the variable has a different value, nothing is changed – that is, a CAS operation is the following, done atomically:

```
if *variable == testValue {
    *variable = newValue
    return true
}
return false
```

Now, you can actually implement a non-blocking mutex:

```
func criticalSection() {
    if locked.CompareAndSwap(false,true) {
    defer locked.Store(false)
    // critical section
    }
}
```

This will enter the critical section only if `locked` is `false`. If that is the case, it will atomically set `locked` to `true` and enter its critical section. Otherwise, it will skip the critical section and continue. Therefore, this can actually be used in place of `Mutex.TryLock`.

Practical uses of atomics

Here are a few examples of using atomics. These are simple race-free uses of atomics in different scenarios.

Counters

Atomics can be used as efficient concurrency-safe counters. The following program creates many goroutines, each of which will add 1 to the shared counter. Another goroutine loops until the counter reaches 10000. Because of the use of atomics here, this program is race-free, and it will always terminate by eventually printing 10000:

```
var count int64

func main() {
    for i := 0; i < 10000; i++ {
        go func() {
            atomic.AddInt64(&count, 1)
            }()
        }
    for {
        v := atomic.LoadInt64(&count)
        fmt.Println(v)
        if v == 10000 {
            break
        }
    }
}
```

Heartbeat and progress meter

Sometimes, a goroutine can become unresponsive or not progress as quickly as necessary. A heartbeat utility and progress meter can be used to observe such goroutines. There are a few ways this can be done – for example, the observed goroutine can use non-blocking sends to announce progress, or it can announce its progress by incrementing a shared variable protected by a mutex. Atomics allow us to implement the shared variable scheme without a mutex. This also has the benefit of being observable by multiple goroutines without additional synchronization.

So, let's define a simple `ProgressMeter` type containing an atomic value:

```
type ProgressMeter struct {
    progress int64
}
```

The following method is used by the observed goroutine to signal its progress. This method simply increments the progress value by 1 atomically:

```
func (pm *ProgressMeter) Progress() {
    atomic.AddInt64(&pm.progress, 1)
}
```

The Get method returns the current value of the progress. Note that the load is atomic; without that, there is a chance of missing atomic additions to the variable:

```
func (pm *ProgressMeter) Get() int64 {
    return atomic.LoadInt64(&pm.progress)
}
```

An important detail in this implementation is that both the Progress() and Get() methods must be atomic. Let's say you wanted to also keep the timestamp when the last progress is recorded. You can add a timestamp variable and use another atomic read/write:

```
type WrongProgressMeter struct {
    progress    int64
    timestamp   int64
}

func (pm *WrongProgressMeter) Progress() {
    atomic.AddInt64(&pm.progress, 1)
    atomic.StoreInt64(&pm.timestamp,
        time.Now().UnixNano())
}

func (pm *WrongProgressMeter) Get() (n int64 ,ts int64) {
    n = atomic.LoadInt64(&pm.progress)
    ts = atomic.LoadInt64(&pm.timestamp)
    return
}
```

This implementation can read an updated progress value with a stale timestamp. The use of atomics guarantees that the write operations are observed in the order they are written, but it does not guarantee the atomicity of ProgressMeter updates. A correct implementation should use Mutex to ensure atomic updates.

Now, let's write a long-running goroutine that uses this progress meter to announce its progress. The following goroutine simply sleeps for 120 milliseconds and records its progress:

```
func longGoroutine(ctx context.Context, pm *ProgressMeter) {
    for {
        select {
        case <-ctx.Done():
            fmt.Println("Canceled")
            return
        default:
        }
        time.Sleep(time.Duration(rand.Intn(120)) *
          time.Millisecond)
        pm.Progress()
    }
}
```

The observer goroutine will expect the observed goroutine to record its progress at least every 100 milliseconds. If that doesn't happen, it will cancel the context to terminate the observed goroutine. It will terminate itself as well. With this setup, the observed goroutine will eventually take longer than 100 milliseconds between two progress announcements; hence, the program should terminate:

```
func observer(ctx context.Context, cancel func(), progress
*ProgressMeter) {
    tick := time.NewTicker(100 * time.Millisecond)
    defer tick.Stop()
    var lastProgress int64
    for {
        select {
        case <-ctx.Done():
            return
        case <-tick.C:
            p := progress.Get()
            if p == lastProgress {
                fmt.Println("No progress since
                  last time, canceling")
                cancel()
                return
            }
```

```
                    fmt.Printf("Progress: %d\n", p)
                    lastProgress = p
                }
            }
    }
```

We wire it up by creating the long-running goroutine and its observer, using a context and a progress meter:

```
func main() {
    var progress ProgressMeter
    ctx, cancel := context.WithCancel(
      context.Background())
    defer cancel()
    wg := sync.WaitGroup{}
    wg.Add(1)
    go func() {
        defer wg.Done()
        longGoroutine(ctx, &progress)
    }()
    go observer(ctx, cancel, &progress)
    wg.Wait()
}
```

Note that we pass the `cancel` function to the observer so that it can send a cancellation message to the observed goroutine. We will look at another way of doing that next.

Cancellations

We have already looked at using the closing of a channel to signal cancellations. Context implementations use this paradigm to signal cancellations and timeouts. A simple cancellation scheme can also be implemented using atomics:

```
func CancelSupport() (cancel func(), isCancelled func() bool) {
    v := atomic.Bool{}
    cancel = func() {
        v.Store(true)
    }
    isCancelled = func() bool {
        return v.Load()
```

```
        }
        return
    }
```

The `CancelSupport` function returns two closures – the `cancel()` function can be called to signal cancellation, and the `isCancelled()` function can be used to check whether a cancellation request has been registered. Both closures share an atomic `bool` value. This can be used as follows:

```
func main() {
    cancel, isCanceled := CancelSupport()
    wg := sync.WaitGroup{}
    wg.Add(1)
    go func() {
        defer wg.Done()
        for {
            time.Sleep(100 * time.Millisecond)
            if isCanceled() {
                fmt.Println("Cancelled")
                return
            }
        }
    }()
    time.AfterFunc(5*time.Second, cancel)
    wg.Wait()
}
```

Detecting change

Let's say you have a shared variable that can be updated from multiple goroutines. You read this variable, perform some computation, and now you want to update it. However, another goroutine may have already modified that variable after you had your copy. Therefore, you want to update this variable only if someone else did not change it. The following code snippet illustrates this using CAS:

```
var sharedValue atomic.Pointer[SomeStruct]

func updateSharedValue() {
    myCopy := sharedValue.Load()
    newCopy := computeNewCopy(*myCopy)
    if sharedValue.CompareAndSwap(myCopy, &newCopy) {
```

```
        fmt.Println("Set value successful")
      } else {
      fmt.Println("Another goroutine modified
        the value")
    }
  }
```

This code is race-prone, so you have to be careful. The `sharedValue.Load()` call atomically returns a pointer to the shared value. If another goroutine modifies the contents of the pointed `*sharedValue` object, then we have a race. This works only when all goroutines atomically get the pointer and make a copy of the underlying data structure. Then, we write the modified copy using CAS, which may fail if another goroutine behaved more quicker.

Summary

In conclusion, you do not *need* atomics to implement correct concurrent algorithms. However, they can be nice to have if you identify a concurrency bottleneck. You can replace some simple mutex-protected updates (such as counters) with atomics, provided you also use atomic reads to read them. You can use CAS operations to detect concurrent modifications, but also note that few concurrent algorithms need that.

In the next chapter, we will look at how we can diagnose problems and troubleshoot them in concurrent programs.

10

Troubleshooting Concurrency Issues

All non-trivial programs have bugs. When you realize that there are some anomalies in a program, starting a debugging session is usually not the first thing you should do. This chapter is about some of the techniques you can use for troubleshooting without using a debugger. You may find, especially when dealing with concurrent programs, that debuggers sometimes do not offer much help and the solution relies on careful code reading, log reading, and understanding stack traces.

In this chapter, we will explore the following:

- How to read stack traces
- How to detect failures using additional code to monitor program behavior and sometimes heal the program
- Debugging anomalies using timeouts and stack traces

Technical Requirements

The source code for this particular chapter is available on GitHub at https://github.com/PacktPublishing/Effective-Concurrency-in-Go/tree/main/chapter10.

Reading stack traces

If you are lucky, your program panics when something goes wrong, and prints out lots of diagnostic information. You would be lucky because if you have the output of a panicked program, you can usually figure out what went wrong by just looking at it together with the source code. So, let's take a look at some stack traces. The first example is a deadlock-prone implementation of the dining philosophers problem, with just two philosophers:

```go
func philosopher(firstFork, secondFork *sync.Mutex) {
for {
    firstFork.Lock()
    secondFork.Lock() // line: 10
    secondFork.Unlock()
    firstFork.Unlock()
  }
}

func main() {
    forks := [2]sync.Mutex{}
    go philosopher(&forks[1], &forks[0]) // line: 18
    go philosopher(&forks[0], &forks[1]) // line: 19
    select {} // line: 20
}
```

This program eventually deadlocks because of the cyclic nature of the nested locks. When that happens, the runtime detects that there are no active goroutines left in the program and prints a stack trace. The stack trace starts with the root cause (in this case, deadlock):

fatal error: all goroutines are asleep - deadlock!

Then it lists all active goroutines, starting from the one that caused the panic. In the case of deadlock, this can be any one of the deadlocked goroutines. The following stack trace starts with the empty select statement in main (line 20). It shows that there is a goroutine waiting for that select statement:

```
goroutine 1 [select (no cases)]:
main.main()
/home/bserdar/github.com/Writing-Concurrent-Programs-in-Go/
chapter10/stacktrace/deadlock/main.go:20 +0xa5
```

The second goroutine stack shows the path the goroutine followed:

```
goroutine 17 [semacquire]:
sync.runtime_SemacquireMutex(0x0?, 0x1?, 0x0?)
/usr/local/go/src/runtime/sema.go:77 +0x25
sync.(*Mutex).lockSlow(0xc00010e000)
/usr/local/go/src/sync/mutex.go:171 +0x165
sync.(*Mutex).Lock(...)
/usr/local/go/src/sync/mutex.go:90
main.philosopher(0xc00010e008, 0xc00010e000)
/home/bserdar/github.com/Writing-Concurrent-Programs-in-Go/
chapter10/stacktrace/deadlock/main.go:10 +0x66
created by main.main
/home/bserdar/github.com/Writing-Concurrent-Programs-in-Go/
chapter10/stacktrace/deadlock/main.go:18 +0x65
```

You can see that the first entry is from the runtime package, the unexported `runtime_SemacquireMutex` function, called with three arguments. The arguments that are displayed with question marks are values that cannot be captured reliably by the runtime because they were passed in registers instead of being pushed onto the stack. We can be quite sure that at least the first argument is not correct because, if you look at the source code printed in the stack trace (`.../go/src/runtime/sema.go:77`), it is the address of an `uint32` value. (If you test this code yourself, the line number may not match what is displayed here. The important point is that you can still check both the Go standard library functions and your functions by looking at the line numbers printed in your environment.) This function is called by `Mutex.lockSlow`. If you check the source code, you will see that `Mutex.lockSlow` does not take any arguments, but the stack trace shows one. That argument is the receiver for the `lockSlow` method, which is the address of the mutex it is called on. So, here we can see that the mutex that is the subject of this call is at address `0xc00010e00`. Moving down to the next entry, we see that this method was called by `Mutex.Lock`. The next entry shows where this `Mutex.Lock` is called in our program: at line 10. This corresponds to the `secondFork.Lock` line. The next entry in the stack trace also shows that this goroutine was created by `main`, at line 18.

Taking note of the arguments passed to the functions, we see that the `main.philosopher` function gets two arguments: the addresses of the two mutexes. Since the `lockSlow` method is passed the mutex at address `0xc00010e000`, we can infer that it is `&forks[0]`. So, this goroutine is blocked while trying to lock `&forks[0]`.

The third goroutine follows a similar path, but this time the philosopher goroutine is started at line 19, corresponding to the second philosopher. Following similar reasoning, you can see that this goroutine is trying to lock mutex at `0xc00010e008`, which is `&forks[1]`:

```
goroutine 18 [semacquire]:
sync.runtime_SemacquireMutex(0x0?, 0x0?, 0x0?)
/usr/local/go/src/runtime/sema.go:77 +0x25
sync.(*Mutex).lockSlow(0xc00010e008)
/usr/local/go/src/sync/mutex.go:171 +0x165
sync.(*Mutex).Lock(...)
/usr/local/go/src/sync/mutex.go:90
main.philosopher(0xc00010e000, 0xc00010e008)
/home/bserdar/github.com/Writing-Concurrent-Programs-in-Go/
chapter10/stacktrace/deadlock/main.go:10 +0x66
created by main.main
/home/bserdar/github.com/Writing-Concurrent-Programs-in-Go/
chapter10/stacktrace/deadlock/main.go:19 +0x9b
```

This stack trace shows where the deadlock occurred. The first goroutine is waiting to lock `&forks[0]`, which implies it already locked `&forks[1]`. The second goroutine is waiting to lock `&forks[1]`, which implies it already locked `&forks[0]`, hence the deadlock.

Now, let's take a look at a more interesting panic. The following program contains a race, and it panics occasionally:

```go
func main() {
    wg := sync.WaitGroup{}
    wg.Add(2)
    ll := list.New()
    // Goroutine that fills the list
    go func() {
        defer wg.Done()
        for i := 0; i < 1000000; i++ {
            ll.PushBack(rand.Int()) // line 18
        }
    }()
    // Goroutine that empties the list
    go func() {
        defer wg.Done()
```

```
        for i := 0; i < 1000000; i++ {
            if ll.Len() > 0 {
                ll.Remove(ll.Front())
            }
        }
    }()
    wg.Wait()
}
```

This program contains two goroutines: one adding elements to the end of a shared linked list, and the other removing elements from the beginning of the list. This program may run to completion every so often, but sometimes it will panic with a stack trace that looks like this:

```
panic: runtime error: invalid memory address or nil pointer
dereference
[signal SIGSEGV: segmentation violation code=0x1 addr=0x0
pc=0x459570]

goroutine 17 [running]:
container/list.(*List).insert(...)
/usr/local/go/src/container/list/list.go:96
container/list.(*List).insertValue(...)
/usr/local/go/src/container/list/list.go:104
container/list.(*List).PushBack(...)
/usr/local/go/src/container/list/list.go:152
main.main.func1()
/home/bserdar/github.com/Writing-Concurrent-Programs-in-Go/
chapter10/stacktrace/listrace/main.go:18 +0x170
created by main.main
/home/bserdar/github.com/Writing-Concurrent-Programs-in-Go/
chapter10/stacktrace/listrace/main.go:15 +0xcd
exit status 2
```

It is a segmentation violation error, which means the program tried to access parts of memory it is not allowed to. In this case, the panic says it is addr=0x0, so the program tried to access the contents of a nil pointer. The stack trace shows how this happened: main.go:18 is where List.PushBack is called. Following the stack trace from bottom to top, we see that List.PushBack calls List.

insertValue, which then calls List.insert. The nil pointer access happened in List.insert, at line 96, whose source code is available as follows:

```
92: func (1 *List) insert(e, at *Element) *Element {
93:     e.prev = at
94:     e.next = at.next
95:     e.prev.next = e
96:     e.next.prev = e   // This is where panic happens
97:     e.list = 1
```

Now, some simple deductive reasoning: line 96 can panic if e is nil, or e.next is nil. Looking at the source, e cannot be nil because otherwise, it would have panicked before line 96. Then, e.next must be nil. Is there a bug in the standard library code then, because the assignment is done without a nil check?

In these situations, it helps to learn a bit more about the underlying code than to just insert a nil check in the code. If you look at the comments in the source code, you'll see that:

```
// To simplify the implementation, internally a list 1 is
//implemented
// as a ring, such that &l.root is both the next element of
//the last
// list element (1.Back()) and the previous element of the
//first list
// element (1.Front()).
```

Since the list is implemented as a ring, the next and prev pointers cannot be nil. Even if there is only one node in the list, the pointers of that node will point to the node itself. So, somewhere else in the code, these must be set to nil. Browsing the source for assignment to nil values, we find the following:

```
func (1 *List) remove(e *Element) {
    e.prev.next = e.next
    e.next.prev = e.prev
    e.next = nil // avoid memory leaks
    e.prev = nil // avoid memory leaks
    e.list = nil
    1.len--
}
```

There it is! When an element is removed from the list, it is detached from the ring by assigning its pointers to `nil`. This behavior suggests a race where `insert` and `remove` are running concurrently. The next pointer of a node is set to `nil` by `List.remove`, but that removed node is used at the argument for `List.insert`, causing the panic. The solution is to create a mutex and move all list operations inside critical sections guarded by that mutex (and not to add `nil` checks.)

As I tried to show here, it is always advisable to fully understand the situation in which a panic happens. Most of the time, this will require you to research and find out the assumptions about the underlying data structures. Like in the previous linked-list example, if the data structure is written so that it cannot have `nil` pointers, then you should not add a `nil` check when you see one, and try to understand why you ended up with a `nil` pointer.

Detecting failures and healing

Most software systems will fail despite the efforts spent on testing them. This suggests there are limits to what can be achieved by testing. These limitations stem from several facts about non-trivial systems. Any non-trivial system interacts with its environment, and it is simply not practical (and in many cases, not possible) to enumerate all possible environments in which the system will run. Also, it is usually possible to test a system to make sure it behaves as expected, but it is much harder to develop tests to make sure the system does not behave unexpectedly. Concurrency adds additional complexities: a program that was successfully tested for a particular scenario may fail for the same scenario when put into production.

In other words, no matter how much you test your programs, all sufficiently complex programs will eventually fail. So, it makes sense to architect systems for graceful failure and quick recovery. Part of this architecture is the infrastructure to detect anomalies and, if possible, heal them. Cloud computing and container technologies provide many tools to detect program failures and orchestration tools to restart them. Other monitoring, alerting, and automated recovery tools are also available for non-cloud-based deployments with traditional executables.

Some of these anomalies are accumulative bugs that grow over time until all the resources are consumed. Memory leaks and goroutine leaks are such failures. If you notice repetitive program restarts with out-of-memory errors, you should search for a memory or goroutine leak. The Go standard library provides the tooling for this:

- Use the `runtime/pprof` package to add profiling support to your program and run it in a controlled environment to replicate the leak. It makes sense to add a flag to enable profiling, so you can turn profiling on and off without recompilation. You can use the CPU or heap profile to determine the source of the leak.

- Use the `net/http/pprof` package to publish profiles over an HTTP server, so you can observe how your program uses memory as it is running.

Sometimes, these anomalies are not bugs but are caused by reliance on other systems. For example, your program may rely on a response from a service that sometimes takes a very long time to return, or that fails to return at all. This is especially problematic if there is no way to cancel that service. Most network-based systems eventually time out, but that timeout can be an unacceptable value for your program. It is also possible that your program calls out to a service or a third-party library that just hangs. A practical solution for this may be to terminate the program gracefully and let the orchestration system start a new instance of the program.

The first problem is detecting failure. Let's come up with a non-trivial realistic example: suppose we have a program that calls a `SlowFunc` function, which sometimes takes a long time to complete. Also, suppose there is no way to cancel that function. But we don't want our program to wait indefinitely for the results of `SlowFunc`, so we develop the following scheme:

- If the call to `SlowFunc` succeeds within a given duration (`CallTimeout`), we return the result.

- If the call to `SlowFunc` lasts longer than `CallTimeout`, we return a `Timeout` error. Since there is no way to cancel `SlowFunc`, it will continue running in a separate goroutine until completion.

- There may be many calls to `SlowFunc` that take a long time, so we want to limit the number of active concurrent calls to a given number. If all the available concurrent calls are waiting for `SlowFunc` to complete, then the function should fail immediately with a `Busy` error.

- If none of the calls from the maximum number of concurrent calls respond within a given timeout (`AlertTimeout`), we raise an alert.

Let's start developing this scheme as a generic type, `Monitor[Req,Rsp any]`, where Req and Rsp are the request and response types, respectively:

```go
type Monitor[Req, Rsp any] struct {
    // The amount of time to wait for the function
    // to return
    CallTimeout  time.Duration
    // The amount of time to wait to raise an alert
    // when all concurrent calls are in progress
    AlertTimeout time.Duration
    // The alert channel
    Alert        chan struct{}
    // The function we are monitoring
    SlowFunc     func(Req) (Rsp, error)
    // This channel keeps track of concurrent calls
    // to SlowFunc
    active  chan struct{}
```

```
      // A signal to this channel means that the active
      // channel is full
      full      chan struct{}
      // When an instance of SlowFunc returns, it will
      // generate a heartBeat
      heartBeat  chan struct{}
}
```

The `Monitor.Call` function calls `Monitor.SlowFunc` by implementing the timeout scheme described previously. This function can return one of three possible values: a valid response with or without an error, a timeout error after `Monitor.CallTimeout`, or a `Busy` error immediately:

```
func (mon *Monitor[Req, Rsp]) Call(ctx context.Context, req
Req) (Rsp, error) {
    var (
        rsp Rsp
        err error
    )
// If the monitor cannot accept a new call, return ErrBusy
//immediately, but also start the alert timer
    select {
    case mon.active <- struct{}{}:
    default:
        // Start the alert timer
        select {
        case mon.active <- struct{}{}:
        case mon.full <- struct{}{}:
            return rsp, ErrBusy
        default:
            return rsp, ErrBusy
        }
    }

// Call the function in a separate goroutine
    complete := make(chan struct{})
    go func() {
        defer func() {
            // Notify the monitor that the function
```

```
            //returned
            <-mon.active
            select {
            case mon.heartBeat <- struct{}{}:
            default:
            }
            // Notify the caller that the call is complete
            close(complete)
        }()
        rsp, err = mon.SlowFunc(req)
    }()
    // Wait for the result or a timeout
    select {
    case <-time.After(mon.CallTimeout):
        return rsp, ErrTimeout
    case <-complete:
        return rsp, err
    }
}
```

Let's dissect this method. Upon entry, the method attempts to send to mon.active. This is a non-blocking send, so it will succeed only if the number of active concurrent calls is smaller than the maximum allowed. If there are already a maximum number of concurrent calls in progress, the default case is chosen, which attempts to send to mon.active again. This is a typical way of emulating priority between channels. Here, mon.active is given priority. If it cannot send to mon.active, then it tries to send to mon.full. This will be enabled only if there is a goroutine waiting to receive from it, and later, it will be clear that this is only possible if mon.active is full but an alert timer has not been started yet. If the timer is started, the goroutine controlling the alert timer will not be listening from this channel, so the default case will be selected. If that happens, the call returns ErrBusy. If the send to mon.full is successful, then this call will be the one that starts the timer, and it will return ErrBusy.

The second part is the actual call to mon.SlowFunc. This is done in a separate goroutine. Since there is no way to cancel mon.SlowFunc, this goroutine will only return when mon.SlowFunc returns. If mon.SlowFunc returns, several things happen: first, a receive from mon.active removes one entry from it, so the monitor can accept another call. Second, a non-blocking send to the mon.heartBeat channel will stop the alert timer. This is a non-blocking send because, as will be shown later, if the alert timer is active, the send to mon.heartBeat will succeed, and if the alert timer is not active, the goroutine is not listening to the mon.heartBeat channel.

In the final section, we wait for the results of mon.SlowFunc. If the complete channel is closed, then we have the results ready and we can return them. If timeout strikes first, then we return ErrTimeout. If mon.SlowFunc returns after this (which will most likely happen), the result is thrown away.

The interesting piece is the monitor goroutine itself. It is embedded in a NewMonitor function:

```
func NewMonitor[Req, Rsp any] (callTimeout time.Duration,
    alertTimeout time.Duration,
    maxActive int,
    call func(Req) (Rsp, error)) *Monitor[Req, Rsp] {
    mon := &Monitor[Req, Rsp]{
        CallTimeout:  callTimeout,
        AlertTimeout: alertTimeout,
        SlowFunc:     call,
        Alert:        make(chan struct{}, 1),
        active:       make(chan struct{}, maxActive),
        Done:         make(chan struct{}),
        full:         make(chan struct{}),
        heartBeat:    make(chan struct{}),
    }


    go func() {
        var timer *time.Timer
        for {
            if timer == nil {
                select {
                case <-mon.full:
                    timer = time.NewTimer(mon.AlertTimeout)
                case <-mon.Done:
                    return
                }
            } else {
                select {
                case <-timer.C:
                    mon.Alert <- struct{}{}
                case <-mon.heartBeat:
                    if !timer.Stop() {
                        <-timer.C
```

```
                        }
                case <-mon.Done:
                    return
                }
                timer = nil
            }
        }
    }()


    return mon
}
```

The goroutine has two states: one when `timer==nil`, and one when it is not. When `timer==nil`, it means that fewer than `maxActive` concurrent calls are in progress, so there is no need for an alert timer. In this state, we listen to the `mon.full` channel. As we saw previously, if one of the calls sends to the `mon.full` channel, the monitor creates a new timer and we enter the second state. In the second state, we listen to the `timer` channel and the `heartBeat` channel (and not the `mon.full` channel, so the non-blocking send is necessary in `mon.Call`). If `mon.heartBeat` comes before the timer, we stop the timer, and set it to `nil`, putting the goroutine in the first state again. If the timer strikes first, we raise an alert.

To use the monitor, you have to initialize it once and call `SlowFunc` through the monitor:

```
// Initialize monitor with 50 millisecond call timeout, 5
//second alert timeout with at most 10 concurrent calls.
//The target function is SlowFunc
var myMonitor = NewMonitor[*Request,*Response](50*time.
Millisecond,5*time.Second,10,SlowFunc)
```

We have to set up a goroutine to handle alerts:

```
go func() {
    for {
        select {
        case <-myMonitor.Alert:
            // Handle alert
        case <-myMonitor.Done:
            return
        }
    }
}()
```

Then, call the target function through the monitor:

```
response, err := myMonitor.Call(request)
```

In this scenario, there really isn't much you can do to recover from the error. When an alert is triggered, you can send an email or a chat message to notify someone, or simply print out a log and terminate, so a fresh new process can be restarted.

Sometimes it makes sense to restart a failing goroutine. When a goroutine becomes unresponsive for a long time, a simple monitor can cancel the goroutine and create a new instance of it:

```go
func restart(done chan struct{}, f func(done, heartBeat chan
struct{}), timeout time.Duration) {
    for {
        funcDone := make(chan struct{})
        heartBeat := make(chan struct{})
        // Start the func in a new goroutine
        go func() {
            f(funcDone, heartBeat)
        }()
        // Expect a heartBeat before the timeout
        timer := time.NewTimer(timeout)
        retry := false
        for !retry {
            select {
            case <-done:
                close(funcDone)
                return
            case <-heartBeat:
                // heartBeat comes before timeout, reset
                //timer
                if !timer.Stop() {
                    <-timer.C
                }
                timer.Reset(timeout)
            case <-timer.C:
                fmt.Println("Timeout, restarting func")
                // Attempt to stop the current function
                close(funcDone)
```

```
                    // Break out of the for-loop so a new
                    //goroutine can start
                    retry = true
                }
            }
        }
    }
    ...
    // Run longRunningFunc with 100 millisecond timeout
    restart(doneCh, longRunningFunc, 100*time.Millisecond)
```

This will restart `longRunningFunc` if it fails to deliver a heartbeat signal every 100 milliseconds. This can be because `longRunningFunc` failed, or because it is waiting for another long-running process that became unresponsive.

Debugging anomalies

Concurrent algorithms have a way of working when observed and failing when not. Many times, a program that runs just fine in a debugger fails mysteriously in production environments. Sometimes, such failures come with a stack trace, and you can track it back to why it happened. But sometimes, failures are much more subtle with no clear indication of what went wrong.

Consider the monitor in the previous section. You might want to find out why `SlowFunc` hangs. You cannot really run it in a debugger and step through the code because you simply have no control over which invocation of the function hangs. But what you can do is print a stack trace when it happens. This is the nature of most anomalies in concurrent programs: you don't know when it is going to happen, but you can usually tell that it did. So, you can print all sorts of diagnostic information to backtrack how the program got there. For instance, you can print the stack trace when the monitor raises an alert:

```
import (
"runtime/pprof"
    ...
)
...
go func() {
    select {
    case <-mon.Alert:
        pprof.Lookup("goroutine").WriteTo(os.Stderr, 1)
    case <-mon.Done:
        return
```

```
        }
    } ()
```

This will give the stack trace for all goroutines at the time the alert is raised, so you can see which goroutines called `SlowFunc` and what it is waiting for.

Detecting a deadlock is easy if the deadlock includes all active goroutines. When the runtime realizes that no goroutine can proceed, it prints a stack trace and terminates. It is not so trivial if at least one goroutine is still alive. A common scenario is a server in which handling a request results in a deadlock that includes only a few goroutines. Since the deadlock does not block all the goroutines, the runtime will never detect this situation, and all the goroutines in the deadlock will leak. If you suspect such a leak, it may make sense to add some instrumentation to diagnose the problem:

```
func timeoutStackTrace(timeout time.Duration) (done func()) {
    completed := make(chan struct{})
    done = func() {
        close(completed)
    }
    go func() {
        select {
        case <-time.After(timeout):
            pprof.Lookup("goroutine").WriteTo(os.Stderr, 1)
        case <-completed:
            return
        }
    } ()
    return
}
```

The `timeoutStackTrace` function waits until the `done` function is called, or timeout happens. If a timeout happens, it prints out the stack trace for all active goroutines, so you can try to find the reason for the timeout. It can be used as follows:

```
func (svc MyService) handler(w http.ResponseWriter, r *http.
Request) {
    // Print tack trace if the call does not complete
    // in 20 seconds
    defer timeoutStackTrace(20*time.Second)()

    ...

}
```

As you can see, if you suspect a problem like a deadlock or unresponsive goroutines, printing out the stack trace after detecting such an occasion could be an effective way for troubleshooting.

Dealing with a race condition is usually harder. The best course of action is usually to develop a unit test replicating the situation in which you suspect a race and run it with the Go race detector (use the -race flag). The race detector adds the necessary instrumentation to the program to validate memory operations and reports a memory race when it detects one. Since it relies on code instrumentation, the race detector can detect races only when they happen. This means that if the race detector reports a race, then there is one, but if it does not report a race, it doesn't mean there isn't one. So, make sure you run your tests for race detection for a while to increase the chance of a race. Many race conditions will manifest themselves as corrupt data structures, like the list example I showed earlier in this chapter. This will require you to do a lot of code reading (including reading the code for the standard library or third-party library code) to identify the source of the problem as a race condition. However, once you realize that you are dealing with a race condition and not another bug, you can insert fmt. Printf or panic statements at critical points in the code when you detect that something that should not happen, happened.

Summary

This chapter showed some techniques that can be useful to troubleshoot concurrent programs. The key idea when dealing with such programs is that you can usually tell when something bad happened only after it happens. So, adding additional code that generates alerts with diagnostic information can be a lifesaver. Many times, this is simply additional logging or Printf statements and panics (a.k.a *Poor Man's Debugger*). Add such code to your programs and keep that code alive in production. Immediate failure is almost always better than incorrect computation.

Further reading

The Go development environment comes with numerous tools for diagnostics:

- The Go unit testing framework: https://pkg.go.dev/testing
- The runtime/pprof package makes your program's internals available to monitoring tools: https://pkg.go.dev/runtime/pprof
- The Go Race Detector ensures your code is race-free: https://go.dev/blog/race-detector
- The Go Profiler is an indispensable tool for analyzing leaks and performance bottlenecks: https://go.dev/blog/pprof

Index

`packtpub.com`

Subscribe to our online digital library for full access to over 7,000 books and videos, as well as industry leading tools to help you plan your personal development and advance your career. For more information, please visit our website.

Why subscribe?

- Spend less time learning and more time coding with practical eBooks and Videos from over 4,000 industry professionals

- Improve your learning with Skill Plans built especially for you

- Get a free eBook or video every month

- Fully searchable for easy access to vital information

- Copy and paste, print, and bookmark content

Did you know that Packt offers eBook versions of every book published, with PDF and ePub files available? You can upgrade to the eBook version at `packtpub.com` and as a print book customer, you are entitled to a discount on the eBook copy. Get in touch with us at `customercare@packtpub.com` for more details.

At `www.packtpub.com`, you can also read a collection of free technical articles, sign up for a range of free newsletters, and receive exclusive discounts and offers on Packt books and eBooks.

Other Books You May Enjoy

If you enjoyed this book, you may be interested in these other books by Packt:

Functional Programming in Go

Dylan Meeus

ISBN: 978-1-80181-116-3

- Gain a deeper understanding of functional programming through practical examples
- Build a solid foundation in core FP concepts and see how they apply to Go code
- Discover how FP can improve the testability of your code base
- Apply functional design patterns for problem solving
- Understand when to choose and not choose FP concepts
- Discover the benefits of functional programming when dealing with concurrent code

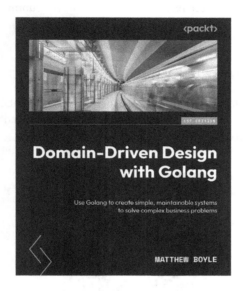

Domain-Driven Design with Golang

Matthew Boyle

ISBN: 978-1-80461-345-0

- Get to grips with domains and the evolution of Domain-driven design
- Work with stakeholders to manage complex business needs
- Gain a clear understanding of bounded context, services, and value objects
- Get up and running with aggregates, factories, repositories, and services
- Find out how to apply DDD to monolithic applications and microservices
- Discover how to implement DDD patterns on distributed systems
- Understand how Test-driven development and Behavior-driven development can work with DDD

Packt is searching for authors like you

If you're interested in becoming an author for Packt, please visit `authors.packtpub.com` and apply today. We have worked with thousands of developers and tech professionals, just like you, to help them share their insight with the global tech community. You can make a general application, apply for a specific hot topic that we are recruiting an author for, or submit your own idea.

Share Your Thoughts

Now you've finished *Concurrency in Golang*, we'd love to hear your thoughts! Scan the QR code below to go straight to the Amazon review page for this book and share your feedback or leave a review on the site that you purchased it from.

`https://packt.link/r/1804619078`

Your review is important to us and the tech community and will help us make sure we're delivering excellent quality content.

Download a free PDF copy of this book

Thanks for purchasing this book!

Do you like to read on the go but are unable to carry your print books everywhere? Is your eBook purchase not compatible with the device of your choice?

Don't worry, now with every Packt book you get a DRM-free PDF version of that book at no cost.

Read anywhere, any place, on any device. Search, copy, and paste code from your favorite technical books directly into your application.

The perks don't stop there, you can get exclusive access to discounts, newsletters, and great free content in your inbox daily

Follow these simple steps to get the benefits:

1. Scan the QR code or visit the link below

https://packt.link/free-ebook/9781804619070

2. Submit your proof of purchase
3. That's it! We'll send your free PDF and other benefits to your email directly

www.ingramcontent.com/pod-product-compliance
Lightning Source LLC
Chambersburg PA
CBHW060557060326
40690CB00017B/3741